Praise for Barry Estabrook

"Smart and important book."

—Sam Sifton, *The New York Times*

"The pleasures of *Tomatoland* are real. They're strong but subtle and sustained. Mr. Estabrook's prose contains a mix of sweetness and acid, like a perfect homegrown tomato itself."

—Dwight Garner, *The New York Times*

"If you care about social justice—or eat tomatoes—read this account of the past, present, and future of a ubiquitous fruit."

—Corby Kummer, TheAtlantic.com

"*Tomatoland* (is) in the tradition of the best muckraking journalism, from Upton Sinclair's *The Jungle* to Eric Schlosser's *Fast Food Nation*."

—Jane Black, *The Washington Post*

"Masterful." —Mark Bittman, *New York Times* Opinion blog

"Eye-opening exposé . . . thought-provoking."

—*Publishers Weekly*

"Estabrook adds some new dimensions to the outrageous . . . story of an industry that touches nearly every one of us living in fast-food nation."

—David Von Drehle, *Time* magazine blog "Swampland"

"*Tomatoland* makes you second-guess your food choices. That Florida red tomato you're eating? Yeah, it's probably gassed to make it that red color, and it also may have been picked by slaves. Not so tasty, eh?" —Carey Polis, The Huffington Post

"Read award-winning journalist Barry Estabrook's *Tomatoland*, and you won't look at a tomato in the same way again . . . Estabrook presents a cogent case for reform, challenging everyone to stand up for what is good not only for the taste buds and the wallet, but also for the soul." —Epicurious.com

"This is the sort of book you want—need—to finish in one or two servings as it will forever changes the way you look at the $6 burger." —*LA Weekly*

"*Tomatoland* has a moral force that I won't soon forget. Estabrook makes it clear that the choice we make between a plastic-tasting supermarket someato and fragrant organic farmer's market tomato . . . says everything about our humanity, and our conception of America as a nation."

—Michele Owens, *Kirkus Book Reviews*

"In the tradition of Michael Pollan and Eric Schlosser, Estabrook gives us the darker side of the fruit we so love. Readers who may not have been turned off the winter version of our collectively favorite fruit will certainly find reason here to pause before making a selection at the supermarket. Choose well, Estabrook reminds us." —*ForeWord Reviews*

"Our favorite fruit may not be quite as innocuous and delicious as it appears." —Salon.com

"Vital information that every conscientious eater—and parents of eaters—ought to know."

—CivilEats.com

"A must read for everyone who eats. I don't care if you are in the commodity cattle business or feed your own family with a small garden. I don't care if you are a policy maker, extension professional, molecular biologist, industrial mogul, minister, teacher, or what have you. *Tomatoland* illustrates how fundamentally bankrupt our current commodity-based, industrial food systems have become and offers a glimmer of hope for a food future that's healthful for all involved. Read it and try not to weep."

—*Grit Magazine*

"Put *Tomatoland* on your reading menu. It will surprise and perhaps enrage you, but its final flavor is hopeful."

—*St. Petersburg Times*

"The buzz about *Tomatoland*, a scathing indictment of South Florida's tomato industry, keeps growing."

—*The Oregonian*

"You can really stop at any point during the narrative and decide that you've bought your last supermarket tomato, but Estabrook is just warming up . . . a brisk read, engrossing as it is enraging."

—TheDailyGreen.com

"Corruption, deception, slavery, chemical and biological warfare, courtroom dramas, undercover sting operations and murder: *Tomatoland* is not your typical book on fruit."

—*Maclean's*

TOMATOLAND

How Modern Industrial Agriculture Destroyed Our Most Alluring Fruit

BARRY ESTABROOK

Andrews McMeel Publishing, LLC

Kansas City • Sydney • London

Andrews McMeel Publishing, LLC
an Andrews McMeel Universal company
1130 Walnut Street, Kansas City, Missouri 64106

www.andrewsmcmeel.com

Portions of this book have appeared in different form in *Gourmet*, *Gastronomica*, *Saveur*, and the *Washington Post*.

11 12 13 14 15 MLY 10 9 8 7 6 5 4 3

ISBN: 978-1-4494-2345-2

Library of Congress Control Number: 2012931540

For the men and women who pick the food we eat

Contents

Acknowledgments

This book would have never been written had Ruth Reichl and John Willoughby at *Gourmet* magazine not had the integrity and courage to print an article about modern-day slavery in a national food magazine. Thanks also to Marisa Robertson-Textor, Christy Harrison, and Adam Houghtaling at *Gourmet* for keeping the story alive online, the facts straight, and the Condé Nast lawyers happy.

My interest in tomato production in Florida was sparked by two terrific magazine articles: "Tomatoes," by Thomas Whiteside (the *New Yorker*, January 24, 1977), and "A Matter of Taste: Who Killed the Flavor in America's Supermarket Tomatoes?" by Craig Canine (*Eating Well*, January/February 1991). That these articles have stood the test of time is both a tribute to the quality of their research and writing and an indication of how little the Florida tomato industry has changed. Three excellent books also inspired and informed me. I am heavily indebted to their authors and heartily recommend their work. *Nobodies* by John Bowe and *The Slave Next Door* by Kevin Bales and Ron Soodalter both examine involuntary servitude in the United States today, and *Ripe* by Arthur Allen provides an engaging, informative portrait of all things tomato. Any writer researching labor abuses in Florida owes an enormous debt to the tireless reporting of Amy Bennett Williams of the *Fort Myers News-Press* and John Lantigua of the *Palm Beach Post*.

Members of the Coalition of Immokalee Workers were generous with their time and tolerance for a curious reporter: Greg Asbed, Lucas Benitez, Emilio Galindo, Laura Germino, Jose Hilario Medel, Leonel Perez, Julia Perkins ("Translator Extraordinaire"), and Geraldo Reyes. I am also grateful for the help I received from Jordan Buckley and Meghan Cohorst of the Student/Farmworker Alliance. Elsewhere in Florida, thanks to Jeannie Economos, Pedro Jesus, Victor Grimaldi, Linda Lee, Greg Schell, Steven Kirk, Barbara Mainster, Tom Beddard, Andrew Yaffa, Yolanda Cisneros, Joseph Procacci, and Reggie Brown.

From academia, Roger Chetelat, Harry Klee, Jay Scott, and Monica Ozores-Hampton were generous with information about tomato breeding and horticulture. Any errors are my own. I would like to stress that information about the effects of agricultural chemicals came from my own research and is in no way attributable to them.

Thanks, Tim Stark and Wayne Miller, for showing me how tomatoes should be grown and letting me ride shotgun. And to Chef Peter Hoffman of Savoy and Back Forty in New York, long a champion of buying local food and great tomatoes.

My agent David Black helped me shape this idea and then placed it with the perfect publisher. Thanks to Kirsty Melville, Chris Schillig, Amy Worley, Tammie Barker, John Carroll, Tim Lynch, Holly Ogden, and Dorothy O'Brien at Andrews McMeel. Much appreciation to Jacinta ("Flying Fingers") Monniere for cleaning up the manuscript and thereby preventing it from being later than it was.

As in everything I do, I had the advantage of the support and shrewd insights of Rux Martin throughout this project, my partner, my teammate, and the most wonderful editor in the world.

introduction

ON THE TOMATO TRAIL

My obituary's headline would have read "Food Writer Killed by Flying Tomato."

On a visit to my parents' condominium in Naples, Florida, I was mindlessly driving along the flat, straight pavement of I-75, when I came up behind one of those gravel trucks that seem to be everywhere in southwest Florida's rush to convert pine woods and cypress stands into gated communities and shopping malls. But as I drew closer, I saw that the tractor trailer was top heavy with what seemed to be green Granny Smith apples. When I pulled out to pass, three of them sailed off the truck, narrowly missing my windshield. Chastened, I eased back into my lane and let the truck get several car lengths ahead. Every time it hit the slightest bump, more of those orbs would tumble off. At the first stoplight, I got a closer look. The shoulder of the road was littered with green tomatoes so plasticine and so identical they could have been stamped out by a machine. Most looked smooth and unblemished. A few had cracks in their skins. Not one was smashed. A ten-foot drop followed by a sixty-mile-per-hour impact with pavement is no big deal to a modern, agribusiness tomato.

If you have ever eaten a fresh tomato from a grocery store or restaurant, chances are good that you have eaten a tomato much like the ones aboard that truck. Although tomatoes are farmed commercially in about twenty states, Florida alone accounts for one-third of the fresh tomatoes raised in the United States, and from October to June, virtually all the fresh-market, field-grown tomatoes in the country come from the Sunshine State, which ships more than one billion pounds to the United States, Canada, and other countries every year. It takes a tough tomato to stand up to the indignity of such industrial-scale farming, so most Florida tomatoes are bred for hardness, picked when still firm and green (the merest trace of pink is taboo), and artificially gassed with ethylene in warehouses until they acquire the rosy-red skin tones of a ripe tomato.

Beauty, in this case, is only skin deep. According to figures compiled by the U.S. Department of Agriculture, Americans bought $5 billion worth of perfectly round, perfectly red, and, in the opinion of many consumers, perfectly tasteless commercially grown fresh tomatoes in 2009—our second most popular vegetable behind lettuce. We buy winter tomatoes, but that doesn't mean we like them. In survey after survey, fresh tomatoes fall at or near the bottom in rankings of consumer satisfaction. No one will ever be able to duplicate the flavor of garden-grown fruits and vegetables at the supermarket (or even the farmers' market), but there's a reason you don't hear consumers bemoaning the taste of supermarket cabbages, onions, or potatoes. Of all the fruits and vegetables we eat, none suffers at the hands of factory farming more than a tomato grown in the wintertime fields of Florida.

Perhaps our taste buds are trying to send us a message. Today's industrial tomatoes are as bereft of nutrition as they are of flavor. According to analyses conducted by the U.S. Department of Agriculture, 100 grams of fresh tomato today has 30 percent less vitamin C, 30 percent less thiamin, 19 percent less niacin, and 62 percent less calcium than it did in the 1960s. But the modern tomato does shame its 1960s counterpart in one area: It contains fourteen times as much sodium.

A couple of winters ago, I bought an assortment of supermarket tomatoes and brought them home for a tasting. I put four on the counter and reached for a cutting board, accidentally nudging one. I was too slow to stop it and watched as it rolled off the counter and fell on our newly refinished pine floor. It hit and traveled for a few feet but incurred no damage. As I retrieved it, my partner came into the kitchen, and I tossed the tomato at her playfully. She shrieked and dodged, and my hardy store-bought tomato struck the floor with the solid thud of a baking potato. I bowled the fruit through the kitchen door, across the dining room, over a wooden threshold, onto the tile floor of the sunroom, where The Tomato That Would Not Die crashed against the door. No damage done.

The best way to experience true tomato taste is to grow your own. Little wonder that tomatoes are by far the most popular vegetable for home gardeners, found in nearly nine out of ten backyard plots. Both The Tomato That Would Not Die and the heirloom Brandywines in my Vermont garden are of the species *Solanum lycopersicum*, and both are red. But the similarity ends there. My Brandywines are downright homely lumpy, deeply creased, and scarred, they look like badly sunburned Rubens derrieres. Nor are they made for travel. More often than not, one will spontaneously split during the twenty-five-yard stroll from garden to kitchen. If not eaten within a day or so after being picked, they develop brownish bruises and begin leaking a watery orange liquid. But that rarely happens. Around our place, Brandywines go fast. They may be ugly. And fragile. Yet there is no better-tasting tomato than a garden-ripe Brandywine. With sweetness and tartness playing off each other perfectly, and juices that burst into your mouth in a surge that forces you to abandon all pretext of good table manners and to slurp, a real tomato's taste is the distilled essence of sun, warm soil, and fine summer days.

Not everyone can grow a garden or head out to a neighborhood farmers' market in search of the ideal tomato. But we all have an alternative to the sad offerings of commercial agriculture. At a lunch spot in the town where I live, a handwritten notation appeared on the blackboard listing the daily specials one June afternoon. "Dear Customers, we will not be putting tomatoes on our sandwiches until we can obtain ones that meet our standards. Thanks." With that small insurrection, the restaurant's proprietor had articulated a philosophy that more of us should embrace: Insist on eating food that meets *our* standards only, not the standards set by corporate agriculture.

Organic, local, seasonal, fresh, sustainable, fair trade—the words have become platitudes that skeptics associate with foodie elitists who can afford to shop at natural food stores and have kitchens that boast $5,000 ranges and larders filled with several varieties of vinegar, extra-virgin olive oil, and "natural" sea salt. It's easy to forget that those oft-repeated words do mean something. Florida's tomato fields provide a stark example of what a food system looks like when all elements of sustainability are violated.

This book began as an attempt to answer what I thought were a couple of simple questions. Why can't (or won't) modern agribusiness deliver a decent tasting tomato? And why can't it grow one with a similar nutritional profile to the tomatoes available to any housewife during the Kennedy administration? My investigations into the mysteries of modern tomato production took me on a circuitous journey from my garden in New England to a research greenhouse at the University of California Davis, to the rocky fields of a struggling produce farmer in Pennsylvania, and to the birthplace of tomatoes in the remote coastal deserts of northern Peru. But I always found myself coming back to where it all started for me—Florida.

So, why can't we walk into a supermarket in December and buy the tomato of our dreams? Part of the reason is that it is essentially against the law. Regulations actually prohibit growers in the southern part of Florida from exporting many of the older tasty tomato varieties because their coloration and shape don't conform to what the all-powerful Florida Tomato Committee says a tomato should look like. The cartel-like Committee exercises Orwellian control over tomato exports from the state, and it decrees that slicing tomatoes shipped from South Florida in the winter must be flawlessly smooth, evenly round, and of a certain size. Taste is not a consideration.

If it were left up to the laws of botany and nature, Florida would be one of the last places in the world where tomatoes grow. Tomato production in the state has everything to do with marketing and nothing to do with biology. Florida is warm when the rest of the East and Midwest—within easy striking distance for a laden produce truck—is cold. But Florida is notoriously humid. Tomatoes' wild ancestors came from the coastal deserts of northern Peru and southern Ecuador, some of the driest places on earth. Taken to Spain, Italy, and southern France in the 1500s, they thrived in the Mediterranean's sunny, rainless summers. They flourish in the dry heat of California, home to the U.S. canned tomato industry, which is completely distinct from the fresh-market tomato industry. Canning tomatoes and fresh tomatoes may as well be apples and oranges. When forced to struggle in the wilting humidity of Florida, tomatoes become vulnerable to all manner of fungal diseases. Hordes of voracious hoppers, beetles, and worms chomp on their roots, stems, leaves, and fruit. And although Florida's sandy soil makes for great beaches, it is devoid of plant nutrients. Florida growers may as well be raising their plants in a sterile hydroponic medium. To get a successful crop, they pump the soil full of chemical fertilizers and can blast the plants with more than one hundred different herbicides and pesticides, including some of the most toxic in agribusiness's arsenal. Workers are exposed to these chemicals on a daily basis. The toll includes eye and respiratory ailments, exposure to known carcinogens, and babies born with horrendous birth defects. Not all the chemicals stay behind in the fields once the tomatoes are harvested. The U.S. Department of Agriculture has found residues of thirty-five pesticides on tomatoes destined for supermarket produce sections.

All of this might have a perverse logic to it if tomato growing were a highly lucrative, healthy business. But it isn't. As large as most of them are, Florida's tomato companies are struggling, always one disaster or disappointing year away from insolvency. Cheap tomatoes from Mexico stream across the border during the winter months. Advances in hydroponic technology have enabled greenhouse tomatoes from Canada and the northern states to eat into Florida's market share during the spring and fall. The industry was nearly dealt a fatal blow in 2008 when it suffered more than $100 million in lost sales after the U.S. Food and Drug Administration erroneously put fresh Florida tomatoes on a shortlist of suspects responsible for a massive salmonella outbreak. Growers lost a similar amount two years later when three-quarters of their plants died during a prolonged freeze. Even at the best of times, commodity tomato farming is a high-stakes gamble. When the replanted fields did eventually ripen after the 2010 cold snap, tomatoes glutted the market and prices dropped so low that it wasn't even worthwhile for growers to harvest their crops. Millions of dollars of perfectly edible tomatoes were left to rot in the fields.

An industrial tomato grower has no control over what he spends on fuel, fertilizer (which requires enormous quantities of natural gas in its manufacture), and pesticides, but he can control what he pays the men and women who plant, tend, and harvest his crops. This has put a steady downward pressure on the earnings of tomato workers. Those cheap tomatoes that fill produce sections 365 days a year, year in and year out, come at a tremendous human cost. Although there have been recent improvements, a person picking tomatoes receives the same basic rate of pay he received thirty years ago. Adjusted for inflation, a harvester's wages have actually dropped by half over the same period. Florida tomato workers, mostly Hispanic migrants, toil without union protection and get neither overtime, benefits, nor medical insurance. They are denied basic legal rights that virtually all other laborers enjoy. Lacking their own vehicles, they have to live near the fields, often paying rural slumlords exorbitant rents to be crammed with ten or a dozen other farmworkers in moldering trailers with neither heat nor air conditioning and which would be condemned outright in any other American jurisdiction.

Paid on a "piece" basis for every bushel-sized basket they gather, tomato pickers are lucky to earn seventy dollars on a good day. But good days are few. Workers can arrive at a field at the appointed time and wait for hours while fog clears or dew dries. If it rains, they don't pick. If a field ripens more slowly than expected, too bad. And if there is a freeze as there was in 2010, weeks can go by without work and without a penny of income. After that freeze, soup kitchens in the state's tomato growing regions (busy enough during "good" times) saw demand exceed capacity. Charitable organizations exhausted their budgets. Unable to pay rent, pickers slept in encampments in the woods. The owners had crop insurance and emergency government aid to offset their losses. The workers had nothing.

And conditions are even worse for some of the men and women in Florida's tomato industry. In the chilling words of Douglas Molloy, chief assistant United States attorney in Fort Myers, South Florida's tomato fields are "ground zero for modern-day slavery." Molloy is not talking about virtual slavery, or near slavery, or slaverylike conditions, but real slavery. In the last fifteen years, Florida law enforcement officials have freed more than one thousand men and women who had been held and forced to work against their will in the fields of Florida, and that represents only the tip of the iceberg. Most instances of slavery go unreported. Workers were "sold" to crew bosses to pay off bogus debts, beaten if they didn't feel like working or were too sick or weak to work, held in chains, pistol whipped, locked at night into shacks in chain-link enclosures patrolled by armed guards. Escapees who got caught were beaten or worse. Corpses of murdered farmworkers were not an uncommon sight in the rivers and canals of South Florida. Even though police have successfully prosecuted seven major slavery cases in the state in the last fifteen years, those brought to justice were low-ranking contract field managers, themselves only one or two shaky rungs up the economic ladder from those they enslaved. The wealthy owners of the vast farms walked away scot-free. They expressed no public regrets, let alone outrage, that such conditions existed on operations they controlled. But we all share the blame. When I asked Molloy if it was safe to assume that a consumer who has eaten a fresh tomato from a grocery store, fast food restaurant, or food-service company in the winter has eaten a fruit picked by the hand of a slave, he corrected my choice of words. "It's not an assumption. It is a fact."

After months of crisscrossing Florida, speaking with growers, trade association executives, owners of tomato-packing companies, lawyers, federal prosecutors, county sheriffs, university horticulturalists, plant breeders, farmworker advocates, soup kitchen managers, field workers, field crew leaders, fair housing advocates, one U. S. senator, and one Mexican peasant who came here seeking a better life for his family only to be held for two years as a slave, I began to see that the Florida tomato industry constitutes a parallel world unto itself, a place where many of the assumptions I had taken for granted about living in the United States are turned on their heads.

In this world, slavery is tolerated, or at best ignored. Labor protections for workers predate the Great Depression. Child labor and minimum wage laws are flouted. Basic antitrust measures do not apply. The most minimal housing standards are not enforced. Spanish is the lingua franca. It has its own banking system made up of storefront paycheck-cashing outfits that charge outrageous commissions to migrants who never stay in one place long enough to open bank accounts. Food is supplied by *tiendas* whose inventory is little different from what you'd find in a dusty village in Chiapas, only much more expensive. An unofficial system of buses and minivans supplies transportation. Pesticides, so toxic to humans and so bad for the environment that they are banned outright for most crops, are routinely sprayed on virtually every Florida tomato field, and in too many cases, sprayed directly on workers, despite federally mandated periods when fields are supposed to remain empty after chemical application. All of this is happening in plain view, but out of sight, only a half-hour's drive from one of the wealthiest areas in the United States with its estate homes, beachfront condominiums, and gated golf communities. Meanwhile, tomatoes, once one of the most alluring fruits in our culinary repertoire, have become hard green balls that can easily survive a fall onto an interstate highway. Gassed to an appealing red, they inspire gastronomic fantasies despite all evidence to the contrary. It's a world we've all made, and one we can fix. Welcome to Tomatoland.

ROOTS

A Chilean soldier was guarding a lonely garrison in the Atacama Desert near the Peruvian border when the American tomato geneticist Roger Chetelat and his field research team arrived. The sentry obligingly provided what should have been straightforward directions to their destination: Follow the road beside the railroad tracks. As an afterthought, he quietly suggested that they be careful not to stray from the road, adding with a knowing nod, "land mines."

Chetelat, an athletic fifty-three-year-old, could be mistaken for a high school gym teacher. In fact he is the director of the prestigious C.M. Rick Tomato Genetics Resource Center at the University of California Davis, the world's foremost repository of wild tomato plants and their seeds. On that day in the desert, Chetelat and his group, which included scientists from the Universidad de Chile in Santiago, had been retracing a trail that had been cold for fifty years, its route filed away in the records of a Chilean herbarium. With luck—lots of it—the stale information might lead them to a few remote clumps of a wild tomato species called *Solanum chilense*. If the team was successful, seeds from those plants, which had never before been collected in that area, would become a valuable addition to the center's collection.

But that was a big "if." First, there were questions concerning the accuracy of the pre-GPS location, given as simply "kilometer 106-108" on the cog railway that switch-backed through the Andes between Chile's port city Arica and La Paz, Bolivia. The notation had been scrawled in the journal of a British collector sometime in the 1950s. Even if the directions were valid, a lot can happen in a half-century to an isolated cluster of plants. Roads get built. Gas pipelines go through. Settlements grow. Fields expand. Animals browse. Facing the distinct possibility that they were on botany's version of the wild goose chase, the researchers had been driving across the desert since dawn and had yet to see anything resembling a wild tomato.

The Atacama Desert makes up the southernmost part of the geographic range of modern tomatoes' wild ancestors, which still grow in parts of western Chile, Peru, and Ecuador (and the Galapagos Islands, home to two errant species). It is a testament to the adaptability of the tomato clan that its members can survive in the Atacama, one of the most inhospitable places on earth. The gravelly, boulder-strewn landscape is fifty times as dry as California's Death Valley. Some parts have not received a drop of rainfall in recorded history. Chetelat has driven across its surface for an entire day without seeing a single living thing. Most of the plants that survive there are low and scrubby and, during the driest months, brown and to all appearances, dead.

Chetelat was further discouraged when the road they had been told to follow diverged from the rail line several kilometers before they reached their goal. Frustrated but still determined, the driver veered onto the tracks, which were still occasionally used, and bounced and jolted along until that became too uncomfortable. Still well short of the marker, the scientists set out on foot, even though it was getting late in the day and no one wanted to bivouac in a semi-militarized no-man's-land. It didn't help that they had not seen a tomato. Until they arrived at kilometer 108, that is. There, just as described, with yellow flowers glowing in the afternoon light, were *S. chilense*, descendants of the plants seen by the 1950s collector. The researchers' reward for a long, uncomfortable session in the field was a handful of seeds not much bigger than grains of sand. Chetelat considered it a good day.

If you enjoy tomatoes, it was a good day for you, too. Their field work in many ways echoes the expeditions of those quirky Victorian naturalists who scoured the globe to add botanical curiosities to their collections. But were it not for the efforts of Chetelat and his predecessors and colleagues at the Rick Center to find and conserve all seventeen species that make up the tomato family, there is a very real possibility that tomato production as we know it today would not exist.

Of all the species that played a part in the great Columbian Exchange—the widespread mingling of plants, animals, and disease organisms between the Eastern and Western hemispheres following the establishment of Spanish colonies in the New World—the tomato surely would have topped the list as the least likely to succeed, never mind to become one of our favorite vegetables. Botanists think that the modern tomato's immediate predecessor is a species called *S. pimpinellifolium* that still grows wild in the coastal deserts and Andean foothills of Ecuador and northern Peru. Inauspicious and easily overlooked, S. *pimpinellifolium* fruits are the size of large garden peas. They are red when ripe and taste like tomatoes, but picking a handful of the diminutive fruits as a snack would take several minutes. Gathering enough for a salad or salsa wouldn't be worth anybody's effort.

Were it not for a few random genetic mutations—mere flukes—chances are that pre-Columbian Americans would never have bothered to domesticate the plants that bore those tiny red berries. Chetelat speculates that some unknown forager or farmer noticed an unusual *S. pimpinellifolium* plant, one that produced larger-than-usual fruits. For reasons that are lost to archeologists, these "deformed" plants were not domesticated in the areas where they grew wild. Researchers have found no evidence of tomatoes depicted on the pottery and tapestries made by natives of what is now northern Peru, which were often elaborately adorned by images of foods important to their diet. Instead, tomatoes were domesticated by Mayan or Mesoamerican farmers somewhere in what is now southern Mexico or northern Central America, more than one thousand miles from the home range of their wild kin. The earliest cultivated tomatoes were of the variety *S. cerasiforme*. Now considered a subspecies of *S. lycopersicum* (the scientific name for domestic tomatoes), *S. cerasiforme* looked and tasted like the cherry tomatoes that are sold in plastic clamshell containers in produce sections and scattered atop fast food salads today. In addition to being small-fruited, *S. cerasiforme* produced long, sprawling vines familiar to any home gardener who has tried to rein in the rampant, weedy growth of varieties like Matt's Wild Cherry, a commonly available type much like the first tomatoes to be cultivated. If you cut any cherry tomato in half, you will notice that it has only two compartments filled with seeds. Some of the early *S. cerasiformes* developed mutations that caused them to produce more than two seed cells. Another mutant strain had a gene that dramatically increased the size of its fruits. Selecting plants that produced larger fruits, or fruits with differing shapes and colors, pre-Columbian farmers created tomatoes that resembled most of the varieties available today. When Hernán Cortés conquered the Aztec city of Tenochtitlan (now Mexico City) in 1521, tomatoes had become an important part of the indigenous diet. Aztec writings even include a dish calling for hot peppers, salt, and *tomatls*—the original recipe for salsa. The Aztecs also had another recipe that required tomatoes, according to the conquistador Bernal Díaz del Castillo. After his troops captured one city, he wrote that the defenders had already prepared a large pot of salt, peppers, and tomatoes in anticipation that victory would provide them with the final ingredient—the flesh of the invading Spaniards.

Spanish explorers wasted no time introducing the beguiling New World fruit to Europe, where it soon established itself. By 1544, just a little more than two decades after their "discovery," the Italian herbalist Pietro Andrea Matthioli published the earliest European reference to tomatoes, calling them *mala aurea*, golden apples. A decade later, Leonhart Fuchs, a German doctor, produced the first known illustration of tomatoes, a colored woodcut showing that the fruit not only arrived in Europe with golden exteriors, as Matthioli's name suggested, but also red skins and in many different shapes and sizes. At first, Europeans viewed tomatoes as merely decorative, but soon they began using them as medicines, most often to treat eye ailments. Introduced to France, tomatoes were called *pommes d'amour* (literally "love apples," but the designation might have been a corruption of the Spanish name, *pome dei Moro*, or Moor's apple). By the end of the sixteenth century, tomatoes had finally entered the diet of southern Europeans. Writing in his 1597 *Herball*, the British barber-surgeon John Gerard reported that "love apples" were eaten in boiled form along with "pepper, salt, and oile" as a sauce, although his assessment of the result would not have made his countrymen salivate in envy of Italian gourmands. "They yield very little nourishment to the bodie, and the same naught and corrupt," Gerard wrote, adding that tomatoes were "of rank and stinking savor." Apparently the Italians disagreed. In 1692 the first cookbook mentioning tomatoes was published in Naples, and *pomodori* were on their way to becoming the signature ingredient of southern Italian cuisine. Although they eschewed eating the "rank and stinking" tomato, the British did begin to use it, not for its culinary merits, but for its curative powers over such maladies as headaches, blockages of the bladder, gout, sciatica, running sores, hot tumors upon the eyes, and vapors in women. The first Britons to dine on these misunderstood love apples were Jews of Portuguese and Spanish ancestry in the mid-1700s.

In the United States, colonists called the love apple by its Mexican name, *tomate*, and in the years following the Revolutionary War grew it and incorporated it widely into their cooking, although some

Americans viewed the fruit as poisonous. They found other uses for it, too. One writer recommended putting fresh vines under blankets as a way to control bedbugs. In the early 1800s, patent medicine hucksters began bottling tomato extract as an elixir, advertising that it would cure ills ranging from constipation to chronic cough to the common cold. Their boosterism sparked a national tomato craze, enabling farmers near big cities to make fortunes. Being prolific, tomatoes provided filling food for hungry soldiers. And being high in acid, they lent themselves to the new technology of risk-free canning. The Union Army left a trail of empty tomato cans in the wake of its campaigns. After the war, the veterans' appetite remained unabated. Expensive, out-of-season fresh tomatoes became status symbols. Tomatoes even made it all the way to the Supreme Court. To protect American farmers from competition from Caribbean growers of fresh winter tomatoes under the Tariff Act of March 3, 1883, the justices in 1893 rewrote the dictionary and decreed that tomatoes were vegetables (they are in fact fruits).

Tomatoes' near-universal popularity in North American kitchens and gardens today can be traced back to the efforts of one man, Alexander W. Livingston, who was born in 1821 in Reynoldsburg, Ohio, just outside Columbus. His career as one of the greatest tomato breeders in history got off to an inauspicious start. In his autobiography, *Livingston and the Tomato*, he recalls: "Well do I remember the first tomato I ever saw. I was ten years old, and was running down one of those old-fashioned lanes, on either side of which was the high rail fence, then so familiar to all Ohio people. Its rosy cheeks lighted up one of these fence-corners, and arrested my youthful attention. I quickly gathered a few of them in my hands and took them to my mother to ask, 'What they were?' As soon as she saw me with them she cried out, 'You must not eat them, my child. They must be poison for even the hogs will not eat them. . . . You may go and put them on the mantle, they are only fit to be seen for their beauty.'"

It's a good thing for tomato lovers that young Alexander ignored his mother's advice. By 1842, Livingston began working for a local seed grower. A decade later he had purchased his own land and turned his attention to developing a tomato that was distinctly better than the gnarly, hollow, and dry fruits that were the norm in the middle of the nineteenth century. After more than a decade of following the accepted wisdom of the era—saving the seeds from the largest and most promising fruits each year and replanting them the next—Livingston revolutionized crop development. Instead of looking at fruits, he sought out whole plants that had desirable traits and crossbred them with varieties that had complementary qualities. He came across a plant that bore large quantities of perfectly round fruits. Unfortunately, they were small, so he crossed and recrossed those plants with large-fruited varieties until, five years after spotting that first smooth-fruited plant in his field, he perfected a variety he called the Paragon.

In addition to being a talented botanist, Livingston had a gift for writing unabashedly hyperbolic advertising copy—a key job requirement for successful seed catalog copywriters to this day. The Paragon "was the first perfectly and uniformly smooth tomato ever introduced to the American public, or, so far as I have ever learned, the first introduced to the world." Giving himself credit where it was due, he wrote, "With these, tomato culture began at once to be one of the great enterprises of this country."

Paragon was just the beginning. Livingston himself went on to personally breed a dozen more successful tomatoes, and by 1937 the U.S. Department of Agriculture estimated that half of the tomato varieties in the country owed a genetic debt to Livingston's early discoveries. It's a testament to the nineteenth-century plant breeder's skills that Paragons can still be found in seed catalogs today. I usually put a few in my garden each summer, my way of paying homage to the Great Man. They may not be the best tomato of the season, nor the most prolific, but, as advertised, Paragons are smooth, round, and juicy. If they have anything to apologize for 140 years after their debut, it's that by being consistently prolific and uniform, they gave rise to the fresh tomato industry whose dubious benefits we reap today.

Florida was a late comer to the commercial tomato game. They were grown there as early as 1870 by two farmers named Parry and Wilson in Alachua in the northern part of the state. Two years later, E. S. Blund was harvesting tomatoes on Sanibel Island in southwest Florida. But Joel Hendrix, a shopkeeper and owner of a commercial steamship dock, as well as a six-acre farm in the settlement of Palmetto, established the commercial model that the Florida tomato industry has followed ever since. On January 6, 1880, Hendrix wrote a letter outlining a business plan that involved exporting green tomatoes from Florida that would ripen on their way to northern markets. He then demonstrated that it could be done successfully by shipping a cargo of the unripe fruit from his field in Manatee County (just south of Tampa and still an important tomato growing area) to New York City. No record remains describing the taste or condition of Hendrix's fruits, which in that era would have endured a bouncy wagon ride over rutted sand trails before being loaded onto one or more steamships and rail cars for the long, often rough journey north. But fresh fruits and vegetables of any sort were rarities in the North at that time, and the Yankees eagerly gobbled up Hendrix's out-of-season tomatoes. Establishing another policy that the Florida industry still follows, Hendrix priced his product inexpensively at a level that the average winter-weary New Yorker could afford. Green, cheap, and off-season continue to be the three mercantile legs upon which Florida's tomato industry stands.

Other farmers followed Hendrix's lead. By 1890, a decade after that first shipment, there were 214 acres of tomatoes growing in Manatee County, according to the U.S. Department of Agriculture. The industry never looked back. Aided by the arrival of the railroad in 1884, land under tomato production in Florida increased to 6,675 acres by 1900, in no small part because the crop thrived in the virgin, disease-free soil. Those carefree days ended abruptly in 1903 when an outbreak of fusarium wilt wiped out tomato plantings. But with plenty of vacant land to exploit, growers simply abandoned diseased fields and cleared new ones, slowly pushing inland from the coast. By 1930, there were twenty-nine thousand acres of tomatoes growing in the Sunshine State.

That was around the time that scientists perfected commercial applications for artificially exposing unripe fruits and vegetables to ethylene, a gas that plants produce naturally as a final step in maturing their fruits. Writing in a 1931 issue of *Industrial and Engineering Chemistry*, E. F. Kohman, a researcher with the National Canners Association, observed that if gassed with ethylene, tomatoes could be picked before they were fully ripened and therefore would withstand handling better than their nongassed counterparts, although he acknowledged: "It should be clearly understood that by no known method of ripening except on the vine can a tomato be produced equal in quality to a tomato fully ripened on the vine." Although Florida farmers wholeheartedly embraced the idea of artificially "degreening" their unripe crops, Kohman's concerns about quality were quickly forgotten.

The person most responsible for ushering in the boom years of the Florida tomato industry was an unsuccessful Cuban lawyer named Fidel Castro. Until the embargo of the early 1960s, Florida tomato farmers faced stiff competition from produce grown on the balmy island to the south. But with a stroke of President Kennedy's pen in 1962, no more Cuban tomatoes could be had in the United States. Florida wasted no time stepping into the void. In 1960 the state grew about 450 million pounds of tomatoes a year. Within five years, the harvest had increased by 60 percent to 720 million pounds; revenues soared seventeenfold from $47 million in 1960 to over $800 million by the 1990s. Tomatoes had become big business.

Max Lipman, a European Jewish immigrant who initially settled in New York City, exemplifies this period of expansion. In 1942, he moved to Florida, where he hoped to make a success with a small vegetable wholesale business, buying from local farmers and shipping their produce to northern customers. Within ten years, he had purchased his own land near Immokalee in the southwestern part of the state and was joined in the business by his three sons and three sons-in-law. Playing off the family name, they called their business Six L's Packing Company. Four generations later, the company is still controlled by the Lipman family. It grows, packs, and ships fifteen million twenty-five-pound boxes of tomatoes a year from a sprawling warehouselike facility on the outskirts of Immokalee. Six L's has captured 12 percent of the Florida tomato market, making it the largest of the dozen or so big growers that now raise and ship virtually all Florida tomatoes. Other large companies in the state, like Pacific Tomato Growers, Procacci Brothers Sales, East Coast Growers and Packers, and DiMare Fresh, share almost identical corporate histories to that of Six L's. Launched by ambitious first- or second-generation immigrants, often from small stalls or push carts in northeastern cities such as Boston, New York, and Philadelphia, they expanded rapidly in the second half of the 1900s to become huge companies that, even after several generations, are still run by descendants of the founding family member.

Today's tomatoes may be big, juicy, and smooth skinned, but on their circuitous journey from the arid hillsides and rocky canyons of coastal South America to our dinner tables, they lost many of the genetic traits that were once critical to their survival. The pea-size *S. pimpinellifolium* and the other wild relatives of modern tomatoes that Chetelat and his team seek out and attempt to preserve are tough, versatile organisms that have evolved disease resistance and tolerance to extreme environmental conditions—genetic traits that researchers can incorporate into cultivated tomatoes, a feeble, inbred lot that, like some royal families and certain overpopularized dog breeds, need all the genetic help they can get.

Drop by nearly any farmers' market on a summer Saturday, and displays of cultivated tomatoes all but scream out the word *diversity*. Small cherry tomatoes, grape tomatoes, pear-shaped salad tomatoes, soft ball–size beefsteak tomatoes the color of fire trucks, plum tomatoes, tomatoes that are ribbed like pumpkins, tomatoes that are as perfectly spherical as a billiard ball, tomatoes that are lobed and lumpy, tomatoes that mature ninety days after being transplanted, tomatoes that require only sixty days, tomatoes that when ripe are red, pink, orange, yellow, purple, green, or any combination thereof. But all that variety is literally only skin deep. Botanists have but one name for all those oddball cultivated tomatoes: *S. lycopersicum.* "Most of the variation you are seeing is from a few genes that control color, shape, and size," said Chetelat. "Other than that, there is very little genetic variation."

The mutant plants that the Mayans domesticated were literally cut off from their ancestral roots, living in isolation more than one thousand miles away from other plants of the same species. As early farmers saved seeds from offspring of the original few plants from year to year, the population became increasingly inbred, a process geneticists call a "bottleneck effect." Chetelat draws an example from human migration to explain this phenomenon. "Imagine a handful of people settling a new continent. They represent only a small part of the genetic diversity that was within the continent they left behind. If there's no more migration, then the diversity is even further reduced by inbreeding." Tomatoes went through a series of such bottlenecks in their prehistoric journey from Peru to Mexico, losing genetic diversity each time, and then went through another series of bottlenecks when conquistadores took them from Mexico to Europe.

The problem of inbreeding is exacerbated in cultivated tomatoes because, unlike their wild brethren who must receive pollen from another plant to produce fertile seeds, they are self-pollinated.

A single domesticated plant can "breed" with itself, and the resulting seeds produce offspring that are basically clones, identical to the parent plant. Not going to the bother of connecting with a mate is a rapid, surefire way to reproduce, but it further decreases genetic diversity, producing generation after generation of plants with the same traits—or lack of them. As a result, all the varieties of cultivated tomatoes that have ever been bred contain less than 5 percent of the genetic material in the overall tomato gene pool. "They seem diverse," said Chetelat. "But at a DNA level they are very similar. If it wasn't for the genes of these wild species, you wouldn't be able to grow tomatoes in a lot of areas. I don't think there is a cultivated plant for which the wild relatives have been more critical."

I met Chetelat and some of the offspring of those Chilean *S. chilense* one cool, misty January afternoon inside a greenhouse belonging to the Rick Center at the University of California Davis, which is named after its founder, the late Charles M. Rick Jr. Charlie, as his associates called him, worked at the facility until shortly before his death in 2002 at age eighty-seven. He was a legendary plant science professor, a pioneer in discovering and preserving the seventeen species of wild tomatoes, and the world's foremost authority on the genetics and evolution of the tomato.

Born in Reading, Pennsylvania, Rick developed a love of horticulture by working in apple orchards as a boy. After getting a PhD from Harvard, he moved to Davis, where he became a professor of plant genetics. Receiving a Guggenheim Fellowship in 1948, he spent a year in Peru, the first of fifteen field trips he would eventually make to South America to collect seven hundred samples of seeds and other genetic material from populations of wild relatives of tomatoes, many of which have since gone extinct in their native habitat and live on today in the collections of the Rick Center. He combined a photographic memory and an indefatigable work ethic with a puckish character and a natural flair for storytelling. Associates say he possessed attributes of Charles Darwin, Mark Twain, and Indiana Jones.

Until the end of his life, he was an easily recognized character on the U. C. Davis campus, mounted on an ancient, thick-tired bicycle with his full white beard, granny glasses, and floppy cotton fishing hats.

More than anything, however, Rick was big hearted and generous in an academic field where professional relationships are often marked by secrecy and competitiveness. It was Rick who instituted the center's policy of giving seeds away. The Rick Center acts like a lending library, nurturing and preserving its 3,600-specimen collection but also making it readily available to scholars and plant breeders worldwide who want to "check out" seeds for their own experiments.

Today, those seeds are kept in a vault that resembles a restaurant's walk-in refrigerator. Chetelat ushered me inside. A roaring compressor kept the air at a chilly forty-five degrees and the humidity at a dry 25 percent. The space was jammed with shelves holding trays that were filled with small manila envelopes containing seeds. Periodically, supplies in the vault are augmented by seeds from plants grown in a greenhouse like the one where I met Chetelat.

Had he not been my guide, I would never have recognized the plants that surrounded us in the Davis greenhouse as being even remotely related to the plump, red tomatoes in the produce section. These varieties were perennials with solid, semiwoody stems, not the one-season wonders of my garden. Some plants were almost moss-like, creeping along the soil like thyme. One, called *S. ochranthum*, climbed until it touched the glass roof twelve feet overhead and then doubled back toward the floor. Chetelat told me that its vines can grow fifty feet tall, completely covering small barns and outbuildings in its native Peru.

Foliage came in all shapes, sizes, textures. Some of the flowers were odorless, but others perfumed the air with the aromas of jasmine and honey. The round, scalloped leaves of *S. pennielli* were covered with what seemed like a bad infestation of white, gnat-size flies. When I touched a leaf, my finger stuck to its surface, a natural version of flypaper that entraps would-be pests (and left me with gummy fingertips). Another plant, *S. juglandifolium*, bore leaves that were tough, wrinkled, and leathery looking, as if they had fallen from an ancient walnut tree. When rubbed, they gave off piney notes. *S. sitiens* could have been mistaken for Italian parsley, except its leaves were stiff and covered in a waxy substance to prevent water loss. When particularly dry, they fold themselves in half to preserve moisture. The plant across the aisle from the parsley look-alike had leaves covered in fine hairs like those on a prepubescent boy's upper lip. Crushed between my fingers, the leaves exuded a powerful piney smell mingled with hints of celery. The fruits hanging from the vines seemed like a haphazard collection of miniature marbles, the biggest not much larger than my little fingernail. They came in an array of colors: black, yellow, purple, green with white stripes, green with a purplish blush. One plant, *S. habrochaites*, a native of the high Andes, bore tiny, furry fruits that smelled like Vicks VapoRub. Chetelat said that the red color we so closely associate with tomatoes was a one-time genetic event carried by a single member of the tomato family, *S. pimpinellifolium*.

Overcome by all this tomato diversity, I plucked a yellowish-green fruit from a plant Chetelat identified as *S. arcanum*. I squeezed it, and a slimy green substance containing dozens of seeds no bigger than pinheads squirted into my palm. I slurped it. The distinct taste of soap assaulted my mouth, followed immediately by a dry, burning bitterness that lasted . . . and lasted. "You're the first visitor here who has been brave enough to eat one of those things," Chetelat said nonchalantly. "Hopefully you won't die." Could this inedible fruit really be a close relative of a plant central to culinary cultures around the world? The zesty yet sweet base of countless soups, sauces, salsas, and condiments? A treat savored unadorned and out of hand on a warm midsummer afternoon?

But the Rick collection is not really about taste. Domestic tomatoes had virtually no innate resistance to common tomato diseases and pests until breeders began crossing them with wild species in the 1940s. "They were fairly a defenseless lot," explained Chetelat. Wild tomatoes, on the other hand, are more robust: "We know of at least forty-four pathogens for which resistance has been found in wild species." Commercial seed companies have bred traits into domestic varieties to combat about half of those pests and diseases. If you buy from a seed catalog, the maladies that a tomato resists are usually represented by a series of letters following the name. These include such notorious plant killers as stem canker, spotted wilt virus, fusarium wilt (the disease that wiped out tomatoes in Florida in the early 1900s), grey leaf spot, nematodes (microscopic worms), tobacco mosaic, and verticillium wilt. "Most of the efforts so far have been focused on disease," said Chetelat. "It's been the first target, because disease resistance often determines whether or not you can grow a tomato, period. But secondly, on a genetic basis, disease resistances tend to be simply inherited. For the most part, you are talking about single, dominant genes that are fairly easy for geneticists and breeders to work with, whereas breeding for something like increased yield or improved flavor involves multiple genes, so it is harder for researchers to get a handle on those issues. Disease resistance is the best justification for this facility."

Chetelat pointed to an example from his own backyard tomato patch. "I happen to live in an area where the soil is infested with nematodes, which are so much of a problem that I cannot grow heirlooms. They immediately get sick, develop root galls, and collapse before I get much fruit. So I grow nematode-resistant hybrids. That resistance comes from a wild species called *S. peruvianum*. It's native to Peru and bears small green fruits that are inedible—they'd probably make you sick to your stomach. But it is resistant to nematodes. That resistance has been bred into hybrids."

And more potential remains untapped. Any grower in the Northeast in the summer of 2009 who had to dig up and either bury or burn every wilted, blackened tomato vine in the garden is familiar with the ravages of *Phytophthora*, commonly known as late blight, the same mold that killed the potato crop in Ireland in the 1840s, causing the Great Famine. Chetelat told me that there are wild species that are quite tolerant to the disease waiting for the attention of a future plant breeder. Currently, the center is working with researchers from India who hope to incorporate from a wild species into domesticated varieties resistance to tomato yellow-leaf curl virus, a devastating disease that limits tomato production around the world.

Wild tomatoes might even help fight disease in humans. Chetelat and his associates have conducted experiments showing that it is feasible to boost the levels of ascorbic acid, lycopene, beta-carotene, and other healthful antioxidants by introducing genes from wild tomatoes into domestic varieties. Because tomatoes and tomato products are a major source of nutrients worldwide, higher antioxidant levels could have enormous health benefits.

The possibilities of using wild traits to improve cultivated tomatoes seem almost limitless. Some wild species grow at chilly altitudes thirty-five hundred meters up in the Andes, tolerating low temperatures that would cause other tomatoes to shrivel and die. Others thrive in humid rainforests. A few can eke out an existence in the desert. They have adapted to scant rain and intense heat, potentially useful for commercial crops in warm, dry areas like California's Central Valley during a time of irregular rainfall and global warming. With advances in the technologies of working with DNA, new areas are opening up for breeders. Better methods will allow scientists to routinely address more complex traits, such as the elusive matter of taste, which is controlled by multiple genes. Chetelat said he viewed it as a time of opportunity.

But, unfortunately, time could be running out for the wild populations upon which future discoveries may depend. Modern agricultural practices and urban sprawl eliminate habitat for wild tomatoes. Herds of goats, llamas, alpacas, and other domestic animals eat and trample them. Even though the Rick Center can produce seed from previously gathered wild specimens, thereby maintaining genetic lines, Chetelat insisted that collections preserved by humans, however carefully, are no substitute for what he calls "in situ" plants, meaning ones that grow in their native environments without human interference. The most obvious difference between the two is that Chetelat and his team grow their wild tomatoes artificially in greenhouses with adequate water, optimum lighting, and no competition. Pests and diseases are chemically controlled. "You're really changing the environment," he said. "And that causes genetic shifts from one generation to the next. It's artificial selection." There are other potential problems. If growers are not careful, pollen can flow between two distinct populations of a species being raised in the same greenhouse. A harried technician might simply mislabel seeds or mishandle them, allowing one variety to mingle with another. "We wouldn't have a problem if we could store seeds forever and if we had an infinite number of seeds to fulfill researchers' requests," said Chetelat. "Of course, that's not the way it works."

So he and his associates must still pack their collection equipment and head back out into the field. Their latest trip to northern Peru in 2009 illustrated the severity of the conservation challenge. When Rick and others had visited the area decades earlier, they made detailed records of where they observed native populations. Chetelat intended to return to the same sites to reexamine those populations.

Since the mid 1990s, intensive sugarcane agriculture has come to low-lying valleys north of Lima, vast monoculture fields carpeting the valley floors from one mountain range to the other. Chetelat talked to farm supervisors and laborers who formerly nibbled the ubiquitous little tomatoes as snacks, as we might pick a few wild blackberries in the fencerow beside a pasture. They told him that when the sugarcane came, accompanied by the usual herbicides and other agricultural chemicals, the tomatoes disappeared.

Everywhere Chetelat went, the story was the same. It was only when his group ventured higher into the mountain valleys, above one thousand meters, where conditions were too rough and available spaces too small for sugar estates, that they began to find wild tomatoes growing in rocky areas and out of cracks in fieldstone walls. "That area is the center of diversity for one of the immediate ancestors of the cultivated tomato," he said. "And now most of those populations are gone."

In an effort to evaluate the situation more thoroughly, Chetelat is hoping to return to the region sometime in the next few years. Despite the modern advances in genetics and DNA mapping, this expedition will be more in the spirit of early plant collectors like Dr. Rick, the center's founder. His goal is to secure funding from the National Science Foundation to work closely with a team of Peruvian graduate students. Like the old-time botanists, they are going to scour the landscape and count individual plants at differing elevations. After a thorough, scientific evaluation of the remaining wild populations, he hopes to convince officials to take steps to preserve these tomatoes before they, too, are bulldozed or blasted with herbicide.

After we left the balmy greenhouse and stepped back into the chilly mist, Chetelat paused before locking the door. Nodding to the vines behind the glass, he said pointedly, "There may be a chance that fifty years from now, someone will find something really important in something that's growing in there. That is what this is all about."

A TOMATO GROWS IN FLORIDA

In Vermont, where I live, as in much of the rest of the United States, a gardener can select pretty much any sunny patch of ground, dig a small hole, put in a tomato seedling, and come back two months later and harvest something. Not necessarily a bumper crop of plump, unblemished fruits, but something. When I met Monica Ozores-Hampton, a vegetable specialist with the University of Florida, I asked her what would happen if I applied the same laissez-faire horticultural practices to a tomato plant in Florida. She shot me a sorrowful, slightly condescending look and replied, "Nothing."

"Nothing?" I asked.

"There would be nothing left of the seedling," she said. "Not a trace. The soil here doesn't have any nitrogen, so it wouldn't have grown at all. The ground holds no moisture, so unless you watered regularly, the plant would certainly die. And, if it somehow survived, insect pests, bacteria, and fungal diseases would destroy it." How can it be, then, that Florida is the source for one-third of the fresh tomatoes Americans eat? How did tomatoes become the Sunshine State's most valuable vegetable crop, accounting for nearly one-third of the total revenue generated?

From a purely botanical and horticultural perspective, you would have to be an idiot to attempt to commercially grow tomatoes in a place like Florida. The seemingly insurmountable challenges start with the soil itself. Or more accurately, the lack of it. Although an area south of Miami has limestone gravel as a growing medium, the majority of the state's tomatoes are raised in sand. Not sandy loam, not sandy soil, but pure sand, no more nutrient rich than the stuff vacationers like to wiggle their toes into on the beaches of Daytona and St. Pete. "A little piece of loam or clay would go a long way," said Ozores-Hampton. "But, hello?—this is just pure sand." In that nearly sterile medium, Florida tomato growers have to practice the equivalent of hydroponic production, only without the greenhouses.

Because of the state's benign weather, disease-causing organisms and insects do not die from the frosts, blizzards, and subzero cold snaps that kill bugs and pathogens every winter in colder growing areas. Basking in the same balmy climate as the state's active retirees, Florida's pests, fungi, and bacteria stay vigorous and healthy year-round, just waiting to attack the next crop of tomatoes. Those endlessly sunny winter days that the state's tourism agency likes to tout in its advertisements also pack high levels of humidity the promoters prefer not to mention. Not so comfortable for humans, humidity is ideal for the growth of blights, wilts, spots, and molds. Hardy native weeds like nut grass ("called that because it can drive you nuts," one farmer told me) can easily out-compete tomato plants. Some weeds are so tough they can punch through plastic mulch laid down to suppress them. The renowned sunshine also means that rain is patchy in the winter months. Because sand retains almost no water, tomatoes have to be irrigated. And even though the daytime skies may be clear and bright, Florida is still in the Northern Hemisphere and days are short in the winter. A tomato growing in South Florida in late December gets only a little over ten and a half hours of sunlight a day, whereas one growing in New Jersey in June gets fifteen hours—nearly 50 percent more. Shorter days mean less vigorous growth. A tomato trying to grow in Florida also experiences debilitating temperature swings that a tomato in California or Ohio never has to face. As many a disappointed vacationer has learned, a stretch of eighty-five-degree beach days can be broken overnight by one of the notorious cold fronts that frequently blow across the state, dropping temperatures into the forties, thirties, and even lower. Add to all that the occasional hurricane that flattens the staked tomatoes and the all-too-frequent January or February frost that leaves thousands of acres of vines blackened and dead, and you have to ask: Why bother trying to grow something as temperamental as a tomato in such a hostile environment?

The answer has nothing to do with horticulture and everything to do with money. Florida just happens to be warm enough for a tomato to survive at a time of year when the easily accessed population centers in the Midwest, Mid-Atlantic, and Northeast, with their hordes of tomato-starved consumers, are frigid, their fields frozen solid under carpets of snow. But for tomatoes to survive long enough to take advantage of that huge potential market, Florida growers have to wage what amounts to total war against the elements. Forget the Hague Convention: We're talking about chemical, biological, and scorched-earth warfare against the forces of nature.

In tomato agribusiness's campaign to defend their crop from the powers that would otherwise destroy a tomato field, Ozores-Hampton, who came from her native Chile to the University of Florida in the mid 1990s to do graduate work, is a key ally. She is an anomaly in the managerial and academic ranks of Florida's tomato industry. She is not only a woman (rare enough) but an energetic, forceful, extremely fit Latina in her forties who spends her workdays toiling in the hot sun of her tomato test plots in rural areas populated by self-proclaimed red necks and crackers, then hops in her car Friday evenings and spins off to her pied-à-terre in Miami's trendy South Beach district. Almost every other nonfieldworker I encountered in the tomato business was white, male, and sporting at the very least a few gray hairs and/or nurturing a developing paunch, the type of guy you'd encounter on a Saturday on a golf course or in a bass boat. "They all call me 'the Tomato Lady,'" Ozores-Hampton said. "Everyone knows who that is."

For someone who does not fit the mold of Tomatoland's Good Ole Boys' Club, Ozores-Hampton has been given tremendous responsibilities. Her territory encompasses most of the state south of Tampa. She is the only university horticulturalist serving an area that has 180,000 acres of vegetables (more than half the state's vegetable acreage). That land generates annual revenues of $1.6 billion to farmers. She is so important to the industry that when her postdoctoral work ended, a group of growers approached the president of the cash-strapped University of Florida and offered to help fund her full-time, tenure-track position for four years. Ozores-Hampton's specialty is soil nutrients. She studies the cycles of plant, soil, and water interaction to determine the optimal level at which fertilizers should be applied so as to maximize production, leaving as little surplus nitrogen and potassium in the soil as possible. Excess fertilizer is a costly waste for farmers and pollutes groundwater, lakes, and rivers that feed such environmentally delicate habitats as the Everglades and Florida Bay. On the March morning that I dropped into her lab/offices in the university's Southwest Florida Research & Education Center near Immokalee, several of Ozores-Hampton's assistants were upending bushel-size plastic tubs of bright green tomatoes onto lab benches. She explained that they were doing evaluations of different varieties that day to see which were the most productive. "It's the cornerstone of agriculture anywhere in the world," she said. "If you don't start with the right varieties, you are not going to succeed."

When I told her I had come hoping to gain an understanding of how the tomatoes that find their way into the nation's supermarkets and fast food outlets are grown, she gestured toward an empty office off to one side of the lab. "You've come to the right place," she said, sitting me down as if I were one of her less promising undergrads.

"If you want to understand how we grow tomatoes, you have to start at the end of the last harvest, which around here is in April." She explained that during the summer, a farmer can do three things with his land. Some growers put in cover crops like sorghum and sudangrass, which incorporate a little organic material into the soil and out-compete any weeds that spring up. Cover crops also disrupt the cycle of pathogens and nematodes (microscopic worms that destroy tomatoes' roots) by putting a different species into rotation with what would otherwise be a monocrop—tomato after tomato after tomato. In addition, cover crops serve to capture and store the nitrogen and other fertilizers left behind after the growing season. Instead of using cover crops, farmers can also simply leave the fields fallow. Weeds come up, and they use the herbicide Roundup to kill the weeds. Still other growers choose to flood the fields, drowning weeds, pathogens, and nematodes.

Whatever route a grower chooses, the land lies fallow for sixty to ninety days. By July, it's time to start preparing for the next season's crop. Step one is all but identical to the first stage of erecting yet another Florida condo development or shopping mall. Heavy equipment removes all traces of vegetation, leaving a perfectly flat, dry rectangle of pristine sand. Before anything else can happen, that sand has to be watered somehow. And fortunately, water is one area where nature gives Florida farmers a break. Although rain can be unpredictable in the winter, the state is awash in ground water and crisscrossed with canals and ditches meant to drain that water from what would otherwise be swampland.

"Good thing, too," Ozores-Hampton said, "or you and I would be sitting underwater right now." The shallow layer of sand sits atop impermeable "hard pan," made up of clay and compacted organic matter. Some farmers use traditional drip irrigation, where hoses with small holes are run between plants to deliver a trickle of water, but most growers in South Florida employ a system unique to the area called "seepage irrigation." They simply pump water into canals and ditches that cross their fields. The water sinks down to the impermeable hard pan, and with nowhere else to go, seeps outward, moistening the sand from below. If a heavy rain falls, the farmer pumps water out of his field back into a larger canal, lowering the water level beneath his plants' roots and maintaining optimum moisture. "To do this type of irrigation, you have to have water, you have to have sandy soil, and you have to have an impermeable layer to keep that water from draining away," said Ozores-Hampton. South Florida is the only agricultural area in the world that has such conditions.

For such a system to be effective, crops must be grown in raised beds that are covered in plastic to slow the evaporation. To fashion these beds, Florida growers use principles familiar to any child who has constructed a sand castle: Too little moisture and your castle disintegrates immediately; too much and it slumps into a pile of shapeless muck. But if the moisture level is just right, you can construct a castle with steep-sided moats, crenellated walls, and smoothly rounded turrets. Once they have the right concentrations of moisture in tomato-field sand, farmers inoculate it with a "bottom mix" of fertilizer containing about 20 percent of the nitrogen and potassium the plant will need. GPS-guided tractors equipped with laser levelers carve straight lines of perfectly flat-topped beds that are about three feet wide and raised eight inches above the trenches that run between them, a process that is called "pulling beds." Another tractor follows closely behind. Its job is to apply the remaining 80 percent of the nitrogen and potassium, called the "hot mix," into two deep grooves that it gouges on the outside edges of the bed. The tomatoes' roots grow to the edge of the deposits of soluble fertilizer and absorb the exact amount they need over the course of the season, growing farther into the deposit as they deplete one area. "The roots are intelligent. They know where to grow," said Ozores-Hampton.

If those roots are going to do their job, however, they must be protected from competitive weeds, disease spores, and especially nematodes, which thrive in Florida. Growers have a ready solution to these problems. They kill everything in the soil. To do so, they fumigate the beds with methyl bromide, one of the most toxic chemicals in conventional agriculture's arsenal. The Pesticide Action Network of North America, a group advocating for stricter controls on pesticide use, rates methyl bromide as a "Bad Actor," a category reserved only for the worst of the worst agricultural poisons. The fumigant can kill humans after brief exposure in small concentrations. Sublethal doses cause disruptions in estrogen production, sterility, birth defects, and other reproductive problems. Banned from most crops, methyl bromide can still be used on strawberries, eggplants, peppers, and tomatoes. The chemical is injected into the newly formed beds, which are immediately sealed beneath a tight wrapper of polyethylene plastic mulch. Then the growers wait while the chemical does its lethal work. Within two weeks, every living organism— every insect, fungus, weed seed, and germ— in the beds is dead. "It's like chemotherapy," said Ozores-Hampton. Once the soil is suitably lifeless, it's time to plant tomatoes.

To do this, yet another tractor traverses the fields. This one tows a contraption that could have come off the drawing board of Rube Goldberg himself. It has six low slung seats behind its rear wheels. Farmworkers sit on the seats with their backsides only inches away from the top of the bed and their legs jutting forward—a position that from a distance looks like a child sitting on a small sled. Near one of each worker's hands is a tray holding hundreds of six-inch-tall, five-week-old tomato seedlings. The machine creeps along the rows, punching holes through the plastic, into which the workers pop the little plants. All of the nutrients the tomatoes will need for the rest of their lives are sealed beneath the plastic, which can be either white to deflect sun in warm parts of the state or black to heat the soil in cooler regions. The plastic impedes weed growth, maintains even moisture levels in the sand, and prevents rain from washing away the fertilizer.

Even though all the plants' basic needs are met, tomato culture remains labor intensive right until the day when a worker manually picks the fruits from the vines. After being put in the ground, a tomato plant will feel the touch of a human hand nearly a dozen more times. Within three or four weeks of planting, the growing seedlings need support. Machines drive wooden stakes into the center of the beds, and workers move down the rows tightly weaving plastic twine around the stakes and between the vines, a process that will be repeated on three or four more occasions during the growing season. Also at this time, the young plants are hand pruned. Using their fingers, workers pluck off suckers, shoots that spring out from the bottom of the plants and from between the main vine and branches. This forces the plant to channel its energies into producing fruit, rather than expending them on sprawling stems and excess foliage. Professional horticulturalists called scouts visit the fields at frequent intervals and carefully survey roughly one out of every twenty acres (forty- to forty-five thousand acres are planted with tomatoes in Florida) checking for insects and diseases. They typically supply the farm manager with twice-weekly reports, and on that basis he decides what insecticides, herbicides, and fungicides to apply.

During the months it spends in the field, a Florida tomato plant can be attacked by at least twenty-seven insect species and twenty-nine diseases. Between ten and fifteen weeds commonly try to outcompete the tomato plant for sunlight and soil nutrients. To combat these pests, a conventional Florida farmer has a fearsome array of more than one hundred chemicals at his disposal. The *Vegetable Production Handbook for Florida 2010–2011*, a 328-page "crop management" manual put out by the University of Florida, lists nineteen available herbicides for tomato production, should nuisance weeds become an issue. These products have macho-sounding trade names such as Aim, Arrow, Touchdown, Cobra, GoalTender, Firestorm, Scythe, and Prowl. Six of the recommended herbicides fall into Pesticide Action Network's rogues' gallery of Bad Actors. They include carcinogens, chemicals that cause damage to the brain and nervous system, chemicals that disrupt the reproductive system and cause birth defects, and chemicals that are so dangerous that even brief exposure can kill a person outright. But that's just the beginning. Tomatoes are notoriously vulnerable to fungal attack in Florida. Growers keep their plants' leaves green and spotless with thirty-one different fungicides, eleven of which are Bad Actors. And should any of the region's numerous and voracious spider mites, potato beetles, armyworms, cabbage loopers, hornworms, tomato fruitworms, flea beetles, whiteflies, thrips, aphids, leafminers, stink bugs, grasshoppers, mealybugs, mole crickets, or blister beetles decide that a farmer's tomatoes would make a good dinner, he can blast them with one of sixty pesticides, seventeen of which make the Bad Actor list. This chemical defense system comes at a cost. According to figures compiled by the Florida Tomato Exchange, an industry group, a grower typically applies more than $2,000 worth of chemical fertilizers and pesticides to every acre of tomatoes (an area about the size of a football field) that he raises during a season. An acre of Florida tomatoes gets hit with five times as much fungicide and six times as much pesticide as an acre of California tomatoes.

A distressing number of those chemicals are still on tomatoes when they reach supermarket produce sections. Using research compiled by the U.S. Food and Drug Administration and the U.S. Department of Agriculture, the Environmental Working Group found that 54 percent of tomato samples contained detectable levels of pesticides, which puts tomatoes in the middle of the forty-nine produce items they surveyed (celery, peaches, strawberries, and apples were the most often contaminated; onions, avocados, and frozen sweet corn the least). U.S. Department of Agriculture studies found traces of thirty-five pesticides on conventionally grown fresh tomatoes: endosulfan, azoxystrobin, chlorothalonil, methamidophos, permethrin trans, permethrin cis, fenpropathrin, trifloxystrobin, o-phenylphenol, pieronyl butoxide, acetamprid, pyrimethanil, boscalid, bifenthrin, dicofol p., thiamethoxam, chlorpyrifos, dicloran, flonicamid, pyriproxyfen, omethoate, pyraclostrobin, famoxadone, clothianidin, cypermethrin, fenhexamid, oxamyl, diazinon, buprofezin, cyazofamid, deltamethrin, acephate, and folpet. It is important to note that residues of these chemicals were below levels considered to be harmful to humans, but in high enough concentrations, three are known or probable carcinogens, six are neurotoxins, fourteen are endocrine disruptors, and three cause reproductive problems and birth defects.

With all the help they can get from their chemical friends, and provided that they are not killed by frosts or blown over by hurricanes, Florida tomatoes are ready for picking after ten to fifteen weeks. Ready for picking, but by no means ripe. An industrial Florida tomato is harvested when it is still hard and green and then taken to a packinghouse, where it is gassed with ethylene until it artificially acquires the appearance of ripeness. But as far back as the 1920s, food scientists had determined that no tomato artificially ripened with ethylene would ever have taste and texture equal to one allowed to ripen naturally. In the field, any fruits that show the slightest blush of pink, let alone red, are left to rot or are scavenged by freelance "pinhookers" who pay a small fee to enter fields that have been harvested and collect fruits showing color to sell to local restaurants and vegetable stands or through pinhookers' markets. It's not that the Florida growers can't pack fully ripe tomatoes. They have done it in the past. But doing so requires frequent harvesting over a long period of time, which is costly. It is more profitable for them and their large fast food and supermarket customers to handle and sell tomatoes that are harvested in two or three passes when they are green, indestructibly hard, and impeccably smooth skinned and have a couple of weeks of shelf life ahead of them. Taste does not enter the equation. "No consumer tastes a tomato in the grocery store before buying it. I have not lost one sale due to taste," one grower said. "People just want something red to put in their salad."

Ozores-Hampton wasn't picking tomatoes on the day I met with her, and besides, her experimental plots pale in size beside the massive operations of big packers. But Joseph Procacci agreed to take me to one of his fields that was being harvested. Procacci is the chairman of

Procacci Brothers Sales Corporation, a huge conglomerate that raises tomatoes and other crops on vast tracks in New Jersey, North Carolina, California, Mexico, Puerto Rico, and in several parts of Florida. You've probably eaten a Procacci tomato. His company grows about 15 percent of the fresh tomatoes consumed in the United States. A spry, avuncular octogenarian with an acute mind and an almost extrasensory ability to read people, Procacci divides his time between the Philadelphia area, where the company is headquartered, and Naples, Florida, where he lives in The Vineyards, a self-contained minicity complete with golf courses, schools, and a hospital. The Vineyards was one of his tomato fields until the mid-1980s, when he and his brother decided that the ground would be more profitable if they put in a crop of McMansions and swimming pools.

Procacci, the son of Italian immigrants, has been in the produce business since 1935. Then eight years old, he came home from school in Camden, New Jersey, to see a loaded vegetable push cart in front of the family house. "Don't come home until you've sold everything," his father ordered. It's a maxim that Procacci has applied every working day of his life over the last three-quarters of a century as he built up his multimillion-dollar produce conglomerate.

When Procacci and I went out to visit some of the six thousand acres of tomato fields that his company farms in South Florida, we traveled in his cream-colored Lincoln Continental. In the hush of the car's air-conditioned interior, we motored east from The Vineyards toward the fields surrounding Immokalee, first on a broad boulevard, then on a more modest two-lane highway, and finally onto a grid of unmarked dirt roads that became thinner and more pothole riddled with each ninety-degree turn, until I was hopelessly lost and began to fear that Procacci was, too. The only indication that we were on the right track was that, all too frequently, an oncoming tractor trailer, top-heavy under a payload of bright green tomatoes, blasted past us in a cloud of sand and dust. The tomatoes looked simply unripe to me, but Procacci assured me that they were what the industry calls "mature greens," ripe in all but coloration, which a day or two of exposure to ethylene gas back at the packinghouse would take care of.

Finally, at a gap in a sugarcane windbreak, Procacci veered into a field. I immediately discovered that Lincolns are called Town Cars instead of Field Cars for good reason. Procacci's behemoth promptly became stuck in the dry, pale gray sand. I had begun to envision an hours-long trek in the hot sun to the nearest outpost of civilization, when a short Hispanic man materialized from between the stalks of cane. Before I could get out to help, he had leaned on the back bumper, bouncing the car and pushing. The tires spun, then grabbed, and we were once again surfing and swaying along a crude tractor path. We turned at another gap in the windbreak and encountered a crew of about three dozen pickers, all Latino, all dressed in long shirts and pants and wearing head protection against the sun—ball caps, straw cowboy hats, sweatshirt hoodies, or simply knotted kerchiefs.

Despite decades of efforts to design machines that can harvest fresh-market tomatoes, as they are called in the business, someone still has to pick each one individually by hand. Fresh-market growers are deeply envious of their brethren who grow tomatoes destined for canning (most of the canning tomatoes sold in the United States are grown in California), which can be harvested by machines. Fortunately for them, the canning varieties are determinant, meaning that their vines stop growing and most of their fruits ripen at the same time. Growers kill the plants with herbicide, then harvest the fruits with machines that winnow the fruits from the desiccated vines and leaves and deposit them into trucks. The occasional dent, gouge, or split doesn't matter, because within hours the tomatoes are in giant vats being cooked in preparation for processing. Fresh tomato varieties, on the other hand, are often indeterminate. The vines keep growing and produce fruit over a long period. Tomatoes on the bottom of a plant will be plump and red, while ones at the top are still tiny, rock hard, and olive green. And there is no machine that can tell the difference.

Procacci and I got out for a closer look, our shoes sinking into the sand. For all their longevity and toughness, mature green tomatoes present a serious problem to growers. Short of cutting one open and examining its seeds and the gooey substance that surrounds them (called locular jelly), it is difficult to tell the difference between a mature green tomato and one that is simply green. When gassed, immature tomatoes obligingly turn the desired shade of red, but they will never develop any flavor—even by the insipid standards of Tomatoland. To get around this problem, field managers examine the crop and then tell pickers on a certain day to take all the tomatoes below, say, the third row of supporting twine and none from higher up. The less mature fruits higher on the vines will be picked by crews that pass through the field again a couple of weeks later. Typically, it takes three passes to bring in a crop. This arbitrary method increases the odds that most of the tomatoes picked are mature greens but provides no guarantee, which is one reason so many industrial tomatoes taste like nothing.

The primary job requirements for a tomato picker are to have fast hands, a back that can withstand being bent double in ninety-degree heat for up to twelve hours a day, and legs that can run over loose sand when carrying a thirty-two-pound bucket called a *cubeta* on one shoulder. Each picker in Procacci's field had an assigned row. Each man—I saw no women—crouched in front of the plants and pawed furiously among the leaves, shoveling a steady stream of green balls into a plastic bucket clamped between his feet and lower calves. How fast he filled that bucket was all that mattered. Workers are reprimanded by field bosses for being too slow, but there is no penalty whatsoever for being too rough. Nor should there be. Nothing, it seems, can crack or bruise a mature green tomato. Upon filling his basket, the picker hoisted it onto one shoulder and ran, his shoes scuffing the sand for traction, to where a truck waited. Lucky workers had to take only a few steps to reach the truck. Less fortunate ones had to sprint for twenty-five yards or more. With a mighty heave, the picker tossed the bucket up to a supervisor on the truck, who unceremoniously dumped the contents into a "gondola," a giant, open trough the size of a billiard table that contains fifteen tons of tomatoes when full. The picker is the last human to lay hands on a tomato until it reaches the supermarket or fast food outlet.

To see the next phase of a commercial tomato's life, Procacci and I followed a loaded truck back toward Naples. Every so often, it hit a pothole or bump, and a few tomatoes would sail off, just as they had on that nearly fateful day when I drove along I-75. What passed for a shoulder on the narrow lane was littered with perfect, green spheres. We followed that roadside trail of green tomatoes for a half hour to the town of Bonita Springs, where Gargiulo, Inc., one of the companies that makes up Procacci's produce empire, has a state-of-the-art packing plant.

Our tomato truck joined a line of similarly laden vehicles in the parking lot at one end of a large, warehouselike building. Even sitting still, tomatoes are subject to abuse. Researchers have found that internal temperatures in tomatoes waiting in the Florida sun to be offloaded can rise to 110 degrees. The optimum temperature for preserving quality is 68 degrees. When our truck's turn came, a worker opened a panel on the side of a gondola. Another directed a high-pressure hose on the tomatoes. The spray, which came out with the force of a fire hose, swooshed out the tomatoes, blasting them into an S-shaped metal trough, where they bobbled through the curves like so many rubber duckies. Drawing closer to the trough, I detected an odor not unlike the one emanating from a public swimming pool. The tomatoes were being bathed in a warm chlorine solution, Procacci explained, to kill any bacteria that might have contaminated them.

Despite such sanitation efforts, the U.S. Food and Drug Administration has implicated fresh tomatoes in at least twelve large multistate food-poisoning outbreaks since 1990, and in several other small local outbreaks. Tomatoes are responsible for 17 percent of all the produce-related foodborne illness incidents in the country, more than any other single vegetable. Between 1998 and 2006, produce grown in Florida sickened over fourteen hundred consumers. Salmonella is of particular concern. Birds, reptiles, and infected fieldworkers are all vectors for salmonella, which can stay alive in the fields and irrigation water for months. The bacteria can get inside the fruits, where it is safe from external attempts to wash it away, through roots, flowers, cuts in stems, and breaks in the fruits' skin. Salmonella can also encase itself in a biofilm, a natural protective sheath it creates for itself on the exterior of tomatoes, rendering washing ineffective and allowing the bacteria to survive packing, storing, and shipping.

Once they had completed the circuit through the chlorination trough, Procacci's tomatoes boarded an escalator, which took them out of the bath water and up toward an opening into the warehouse. Inside, the clatter of machinery was so loud that the verbal component of Procacci's tour was reduced to gesticulations, mime, and the occasional shouted phrase. The incoming tomatoes rolled onto a conveyor belt with slats spaced about two inches apart. The smallest fruits fell through the cracks. They were destined for either cattle feed or compost piles. Those large enough to run this initial gauntlet traveled along a tangled snarl of conveyor belts moving so fast that individual fruits became a single blurred green stream. The stream flowed under a roaring machine that blow-dried the tomatoes and then another hissing device that misted them with mineral oil to prevent spoilage and give them an appetizing sheen. They emerged into a large room where rows of black and Hispanic women called "graders" stood shoulder-to-shoulder on either side of conveyor belts, their hands flying as they discarded fruits that were damaged, imperfectly round, or God forbid, showing any hint of ripeness. Once past the graders, the tomatoes rolled onto belts with round holes that sorted them according to size, small ones dropping through the first, smaller holes, the larger ones continuing until they reached holes with a large enough diameter to allow them to fall through. Another machine automatically placed flat cartons at the end of the lines. Once a box was mechanically filled with its twenty-five-pound quota, another machine slapped on a lid and sent the full cartons toward a final machine that stacked the boxes on wooden pallets, ten to a layer, eight layers high, a ton of tomatoes in all. Once a pallet was full, it was shrink-wrapped in clear cellophane and a lift truck whisked it off to a warehouse, where its contents would be gassed, or as the tomato industry prefers to call it, degreened.

Procacci ushered me into one of the tomato gas chambers, a room about the size of a small grade school gymnasium lined with rows of pallets stacked four high. The quiet was a relief after the eardrum-shattering noise of the packing area. Procacci explained that the tomatoes surrounding and towering over us were being exposed to low concentrations of ethylene, the same gas that tomato plants produce naturally when the time comes for fruit to ripen. I inhaled deeply. The slight sweetish odor was overpowered by the distinct smell of cardboard, but I couldn't detect any hint of gas. Procacci said that the pallets would stay in the warehouse for a few days—longer if their eventual destination was near the warehouse, shorter if they had to face a days-long journey to a distribution center in the North and eventually to a supermarket or institutional dining facility near my home.

The system may work well for big tomato growers and their corporate customers, but two groups come out on the short end of the industrial tomato bargain. Consumers occasionally get a tomato that makes them ill. And they are almost always seduced into buying by the beautiful red exteriors and then—in the produce aisle's version of bait-and-switch—they are rewarded with a mealy mouthful stripped of nutrients and devoid of flavor. "A total gastronomic loss," wrote James Beard in his book *Beard on Food*, published in 1974 but still true today.

The biggest losers in Tomatoland's hell-bent race to produce cheap commodity fruits are the men and women whose labor produces the food we eat. Day in and day out, they enter those poisoned fields and expose themselves to a witch's brew of toxic chemicals. After my tour with Procacci, I met some of those workers I'd seen bent over in the fields. Their horror stories turned my stomach—a total gastronomic loss in the fullest sense.

CHEMICAL WARFARE

Tower Cabins is a labor camp consisting of about thirty drab wooden shacks and a few deteriorating trailers crammed together behind an unpainted wooden fence just south of Immokalee (pronounced like broccoli), a city in the heart of southwest Florida's tomato-growing region. The community of poor migrant laborers is dreary at the best of times, but just before Christmas a few years ago, there were reasons for joy. Three women, all neighbors, were expecting children within seven weeks of each other.

But in the lives of tomato workers, there is a fine line between hope and tragedy. The first baby, the son of twenty-year-old Abraham Candelario and his nineteen-year-old wife, Francisca Herrera, arrived on December 17. They named the child Carlos. Carlitos, as they called him, was born with an extremely rare condition called tetra-amelia syndrome, which left him with neither arms nor legs. About six weeks later, a few cabins away, Jesus Navarrete was born to Sostenes Maceda. Jesus had Pierre Robin Sequence, a deformity of the lower jaw. As a result, his tongue was in constant danger of falling back into his throat, putting him at risk of choking to death. The baby had to be fed through a plastic tube. Two days after Jesus was born,

Maria Meza gave birth to Jorge. He had one ear, no nose, a cleft palate, one kidney, no anus, and no visible sexual organs. A couple hours later, following a detailed examination, the doctors determined that Jorge was in fact a girl. Her parents renamed her Violeta. Her birth defects were so severe that she survived for only three days.

In addition to living within one hundred yards of each other, Herrera, Maceda, and Meza had one other thing in common. They all worked for the same company, Ag-Mart Produce, Inc., and in the same vast tomato field. Consumers know Ag-Mart mainly through its trademarked UglyRipe heirloom-style tomatoes and Santa Sweets grape tomatoes, sold in plastic clamshell containers adorned with three smiling, dancing tomato characters named Tom, Matt, and Otto. "Kids love to snack on this nutritious treat," says the company's advertising.

From the rows of tomatoes where the women were working during the time they became pregnant, the view was not so cheery. A sign at the entry warned that the field had been sprayed by no fewer than thirty-one different chemicals during the growing season. Many of them were rated "highly toxic," and at least three, the herbicide metribuzin, the fungicide mancozeb, and the insecticide avermectin, are known to be "developmental and reproductive toxins," according to Pesticide Action Network. They are teratogenic, meaning they can cause birth defects. If they are used, the U.S. Environmental Protection Agency mandates "restricted-entry intervals" (REIs in the jargon of chemical agriculture), the time that must elapse between when pesticides are applied and when workers can go into the fields. In all three cases, the women said they were ordered to pick the fruit in violation of REI regulations.

"When you work on the plants, you smell the chemicals," said Herrera, the mother of limbless Carlitos. Subsequent investigations showed that Herrera worked in fields that recently had been sprayed with mancozeb twenty-four to thirty-six days after conception, the stages where a child begins to develop neurologically and physically.

Meza recalled: "It has happened to me many times that when you are working and the chemical has dried and turned to dust that you breathe it." Although regulations require that handlers of many of these pesticides use protective eyewear, chemical-resistant gloves, rubber aprons, and vapor respirators, the three pregnant women said they had not been warned of the possible dangers of being exposed to the chemicals. They wore no protective gear, unless you count their futile attempts to avoid inhalation by covering their mouths with bandanas. Herrera said she felt sick the entire time she worked in the field. She described being coated in pesticides and suffering from dizziness, nausea, vomiting, and lightheadedness. Her eyes and nose felt as though they were burning. She developed rashes and open sores.

Giving up work was not an option. Herrera said that her boss, a subcontractor to Ag-Mart, told her if she did not work, she would be kicked out of the room that he was providing. Ironically, the impending arrival of her first child made it all the more important for her and her husband to have a place to live. She worked in the fields from preconception, through the early stages of gestation, right up until her seventh month of pregnancy, only a few weeks before Carlitos's slightly premature arrival. Even after quitting the fields, she continued to hand wash the chemical-soaked clothes of her husband and her brother, Epifanio.

Jesus's jaw deformity proved not to be as dangerous as first thought, and doctors told his mother that the baby's condition would likely improve as he grew older. Violetta's parents had to mourn the death of their child. But after the birth of Carlitos, Herrera and Candelario's problems intensified. The end of the winter picking season in Florida was approaching, and the family would have to migrate north to find work. But Carlitos needed constant medical attention, which he was receiving through a local agency, the Children's Medical Services of Lee County. Even though he was an American citizen by birth, his parents were Mexican and had no documentation. Deportation was a real possibility.

Things took a turn for the worse when at three months of age the baby developed respiratory problems that made it difficult for him to breathe. He had to be flown from a hospital near Immokalee to Miami Children's Hospital. Lacking a car, Herrera and Candelario had to rely on rides from social workers to make the journey across the state, trips that took two and a half hours one way and could be undertaken only on days when Candelario was not required in the fields, where he still had to work to pay the rent. "There was nothing we could do for our little boy," said Candelario.

One of the social workers helping Carlitos's parents realized that the family faced an insurmountable financial burden and needed legal help. The social worker contacted a local lawyer, who confessed that he would have been completely over his head with such a complex case. He did, however, have a colleague who specialized in catastrophic personal injury, product liability, and medical malpractice litigation. He picked up the telephone and put in a call to Andrew Yaffa, a partner in the firm Grossman Roth, which has offices in Miami, Fort Lauderdale, Boca Raton, Sarasota, and Key West. Although they had no way of knowing it, Abraham Candelario, Francisca Herrera, and Carlitos had just caught what might have been the first break they had ever received in their hardscrabble lives. If you are injured in a car accident, hurt on the job, or the victim of a negligent physician, you could do no better than getting Andrew Yaffa to represent you.

As soon as I met him, I understood why Andrew Yaffa became such a successful lawyer. The day I visited, he was working out of the boardroom in his firm's Boca Raton office. "I live out of a FedEx box," he said. "I serve every office we have." That afternoon he had taken over the conference room table. File folders were strewn here and there. His laptop was open. His expensive suit coat was folded over the back of a chair, and his tie was loose. Every few minutes when a cell phone on the table warbled, he glanced at its caller ID and replaced it without missing a conversational beat.

In his early forties at the time of Carlitos's birth in 2004, Yaffa is widely recognized as one of the top lawyers in the state. He has won many multimillion dollar settlements in cases tried before some of Florida's toughest judges. One of Yaffa's competitors in Florida described him to me in an e-mail as "a great lawyer . . . solid person . . . integrity . . . partner in a fabulous law firm . . . creative . . . innovative . . . bright . . . ethical . . . the works!"

Yaffa is tall and has the sort of telegenic good looks that would make him a shoo-in to play the role of the leading man if someone ever does a movie version of his life as a crusading attorney. His short dark hair is brushed back and moussed neatly in place, and I caught the merest whiff of cologne. His handsome face is tempered by a kind of Midwestern earnestness. (He's actually a Virginia native.) Yaffa establishes an instant rapport, speaking with a soft, unwavering voice. When I asked him why he chose to take on such a long shot case as that of Carlitos Candelario, he eyed me the way he might stare at an uncooperative witness and said, "I see a lot in my work. But when I see a child or a family that has been harmed and in distress, I don't need a whole lot more motivation than that."

Initially, Yaffa could hardly believe what his friend had told him. He needed to see for himself and to talk to the child's parents. Were they people who would come across as credible? Would a jury relate to them? Would they even want his help? Leaving behind his usual car, a new BMW, to avoid drawing attention to himself, he got in the road-weary Chevy Suburban reserved for weekend fishing outings and trips to the beach with his kids and drove from his Miami office across miles of uninhabited saw grass prairies in the Everglades to the shabby two-bedroom trailer that the young couple and their tragically deformed child shared with seven other migrants. When Yaffa knocked on the door, Herrera answered. He was struck by the fact that the petite, round-faced woman was barely older than a child herself. All the men who lived in the trailer were in the fields. Carlitos was propped up in a baby seat. Strips of drying meat hung from a clothesline stretched across the living room, and the humid air was rank and pungent. Flies buzzed everywhere. When Carlitos began fussing, Herrera took the six-month-old baby out of the seat and laid him on the floor. An orphaned puppy that the trailer's residents had adopted came bouncing around, and the child watched it, smiling and cooing. The puppy yipped, pounced, and started nipping at the baby. Carlitos began to scream, and Herrera rushed to pick him up. Yaffa was powerfully affected. The child, who did not even have the ability to flick away a fly or push back against a puppy, faced a lifetime of need. "The pesticides got into her system and affected this child that was forming and lo and behold, he ends up being born with no arms and no legs," he told me.

Speaking in Spanish, he tried to draw out Herrera, who spoke very little Spanish herself. As is the case of many migrant farmworkers, her first language and the one she was most comfortable communicating in was a native Indian dialect. Yaffa explained that a social worker had contacted him, and he was there for one reason—to help her. He told Herrera that there was no pressure for her to work with him. As is the norm for lawyers in his field, he would bear all the legal expenses himself and be paid only by taking a percentage of anything they won.

When Herrera finally nodded her head, Yaffa vowed that he would do everything in his power to help his new client. But even a lawyer of his track record and courtroom acumen had his work cut out for him. Because of all the nearly infinite variables—heredity, exposure to chemicals at other job sites, possible smoking or drug abuse, environmental factors—cases linking pesticide exposure to birth defects are notoriously hard to prove.

Instead of pursuing the conventional approach by trying to determine the chemical that caused the damage and suing the company that made it, Yaffa decided to do something he had never done: He would try to get compensation from the corporate farm where Herrera had worked. In essence, he would try the entire modern agricultural industry and the chemical-based philosophy on which it is founded.

Florida calls itself the Sunshine State. A more accurate moniker might be the Pesticide State. In terms of raw quantities, Florida is awash in toxic chemicals, even compared to other states where agriculture is a lynchpin of the economy. In California, for instance, where the number of acres dedicated to fresh tomato production was virtually the same as in Florida, slightly fewer than one million pounds of insecticides, fungicides, and herbicides were used on those crops in 2006. During the same year, Florida's tomato farmers applied eight times as much: nearly eight million pounds.

Despite all those chemicals being spread across its fields, Florida has an abysmal record when it comes to protecting its farmworkers, employing only about fifty inspectors (ten were hastily added following the media furor surrounding the birth of the deformed Immokalee babies) compared to California's 350. Florida officials take a what-we-don't-know-won't-hurt-us approach to enforcing pesticide application laws and recording instances of farmworkers being exposed to chemicals while on the job. In 2006, a typical year, the Florida Department of Health reported that there were only two definite or probable cases of harmful pesticide exposure among its migrant and seasonal agricultural workforce of roughly 400,000 men and women. Compare that to California, where two hundred cases were recorded. Although California has three times as many laborers on its farms as Florida, that is not enough to account for the hundredfold difference of reported cases between the two states. According to Paula DiGrigoli of the Collier County Health Department, whose district includes the Immokalee area—where tens of thousands of laborers work in citrus and tomato fields each winter—between 2006 and mid-2010 there were no reported cases of pesticide exposure in her region. The likely reason is that regulators in Florida simply looked the other way.

In both Florida and California, physicians are required to report cases of pesticide poisoning. But in Florida that law is unenforced and ignored. The director of the hospital where the three deformed Immokalee babies were born said that he was not even aware that such a law was on the books. He was by no means alone. In 2000, Together for Agricultural Safety, a group working to reduce pesticide exposure, revealed that most Florida health care workers did not know that they were supposed to report cases of suspected and confirmed pesticide poisoning. California, meanwhile, receives thousands of notifications every year. The reason for the discrepancy between the two states cannot be explained by the relative honesty of doctors in the two states nor even by Florida's less-than-robust enforcement policies. In California, a doctor who is treating someone exposed to pesticides must file a report to the state in order to be paid. Because of differences in workers compensation policies, Florida doctors do not have to file similar reports to get reimbursed. With no incentive, it's easy to see how an overworked Florida practitioner might not get around to filling out yet another government form.

When instances of pesticide misuse do surface in Florida, penalties are rarely levied against the farms that are responsible. Between 1993 and 2003, less than 8 percent of the violations of pesticide regulations in Florida resulted in fines. Even when they did, the amounts were miniscule enough for farmers to consider them as just part of the cost of doing business. After the birth of the deformed babies in Immokalee, the state government did launch an investigation into the use of pesticides by Ag-Mart. Agents leveled eighty-eight counts against the firm, fining it $111,200. Ag-Mart denied wrongdoing, and a judge later dismissed seventy-five of the charges (ten had been resolved earlier) and reduced the fine to $8,400.

In Florida, the job of enforcing the U.S. Environmental Protection Agency Worker Protection Standard falls to the Florida Department of Agriculture and Consumer Services, whose mandate is also to promote Florida's agricultural sector. Those standards consist mainly of such common sense requirements as conducting pesticide training programs, delaying workers' entry into sprayed fields for the legally required period after applications, insisting that workers use protective equipment under certain circumstances, posting warning signs around sprayed fields, and immediately transporting workers who have been exposed to chemicals to medical facilities—hardly unreasonable given that America's agricultural workers are more likely to be poisoned by pesticides on the job than those in any other occupation.

A scathing portrait of the Florida Department of Agriculture and Consumer Services's poor or nonexistent pesticide-abuse investigation and enforcement program emerges in the conclusions of Shelly Davis and Rebecca Schleifer of the Farmworker Justice Fund in Washington, DC:

- The state repeatedly failed to find a causal connection between pesticide exposure and injuries suffered by farmworkers. In only two instances out of the forty-six complaints the researchers investigated over a five-year period did it conclude that pesticide exposure had led to worker injury. One of these would seem to have been an open-and-shut case: One worker passed out after applying an extremely toxic pesticide. In the other case, a grower admitted placing unprotected workers in a pesticide-treated field during the quarantine period.
- The state found regulatory violations in thirty-one instances but issued only two fines.
- The state failed to adequately investigate poisoning complaints even when a farmworker was seriously injured or killed. It systematically failed to interview coworkers or other eyewitnesses outside of the presence of supervisory personnel and with adequate translators, failed to obtain relevant medical records, routinely accepted uncorroborated employer claims of compliance, used checklists as a substitute for thorough on-site inspection, and ignored evidence of employer retaliation.

The authors concluded, "Only a completely revamped, enforcement-oriented system can assure farmworkers that their health will be protected and that compliance with the Worker Protection Standard will be achieved." A labor lawyer summed the situation up more colloquially as "a classic case of the fox guarding the henhouse."

But even a completely revamped system won't deal with one of the biggest problems with enforcing pesticide regulations in the state: The vast majority of the workers who are exposed to pesticides on the job do not report the incidents. Lacking medical insurance, pickers are often reluctant to go to doctors because of the cost, and without legal documentation, they go to great lengths to avoid speaking to government officials. Ignorance also plays a role. Workers often are not taught how to identify the initial symptoms of pesticide exposure, which can be similar to those of a chest cold, the stomach flu, a hangover, or a rash. Others are embarrassed by not being macho enough. "Come on, you're a man. You can take it," one boss told a worker I spoke with after he complained of pains in his chest. And often, the pickers are threatened with dismissal by the labor contractors if they report being sprayed or take time off from work to recover.

Guadalupe Gonzales III learned all about that. On September 12, 2005, at about seven o'clock in the morning, Gonzales, a twenty-nine-year-old laborer, arrived at work on a farm operated by Thomas Produce Company. Based in Boca Raton, Thomas farmed thirteen thousand acres, making it one of the biggest players in the Florida produce business at the time. Like many Florida farmworkers, Gonzales did not report directly to the managers of the large packing company but to a contractor, or "crew boss," named Raul Humberto Ruiz. Gonzales's job that day was to apply methyl bromide, the potent soil fumigant. In humans, methyl bromide can damage the lungs, throat, eyes, skin, kidneys, and central nervous system and has been linked to birth defects and cancer. It can trigger pneumonia, heart problems, and in some cases, death.

The U.S. Environmental Protection Agency classifies methyl bromide as a "Category I Acute Toxin," the most deadly category. In addition to warning users to wear long pants and long-sleeved shirts, the label instructions on the chemical's container read, "All persons working with this fumigant must be knowledgeable about the hazards, and trained in the use of the required respirator equipment and detector devices, emergency procedures and proper use of the equipment."

Gonzales, who had been on the job for only two weeks, claimed that he had received no training. In an affidavit, he noted that he was not even told the name of the pesticide he was going to be working with that morning, and he was wearing a short-sleeved shirt. By eleven o'clock, he detected a strong smell. His head started aching, his eyes stung, and his chest was wracked with a severe pain. But when he reported his symptoms to his crew leader, instead of being taken to a medical facility, Gonzales was told to sit in a pickup truck with the air conditioning blasting. He felt better after a few minutes and returned to work, but that evening his condition worsened, and he went to the emergency room at the Hendry Regional Medical Center, where he received treatment. When he went back to the fields two days later, his symptoms returned, so he filed for worker's compensation. Ultimately, the state fined the company $5,000 for pesticide-handling violations, a sum that was reduced to $3,500 on appeal, a slap on the wrist for such a large operation. According to Gonzales, the second time he reported symptoms of chemical poisoning, his boss said, "If you keep feeling bad, I can't keep you as part of the crew anymore," and fired him. Ruiz has denied the claim, but whatever the reason, the thanks that Gonzales got for following official protocol was that he lost his job.

When growers supply workers with the required pesticide training sessions, the results can be almost farcical. I met Victor Grimaldi one morning in the Immokalee offices of the Farmworker Association of Florida, where he is a member of the board of directors. Among other activities, the Farmworker Association tries to educate pickers about the potential dangers of handling pesticides. Grimaldi, a small graying man, who spent twenty years in the fields before retiring, recalled his initial encounter with pesticide-handling instructions. Speaking through a translator, he told me, "I was taken into the office, and the first thing the boss said was, 'Sign this!' It was a document written in English, which I don't read or speak, but I needed work, so what was I going to say?" Grimaldi was then shown a pesticide-handling safety video—also in English—but from the graphics he was able to understand a few snippets of what was going on. Although training is now conducted in Spanish, many workers who speak Mixtec and other Amerindian dialects cannot understand that language, either. Once "trained," Grimaldi was given a backpack-mounted tank full of pesticide, dispatched into a field, and told to start spraying a row of tomatoes. As he worked along, he came near a group of pickers. He abandoned the row they were working in and proceeded to the next. The crew boss came up and said, "Why did you move?" Grimaldi answered that he had just seen a video showing that spraying near people was against the law. "I'm the law out here," the subcontractor said, and ordered Grimaldi back to the row with the pickers.

Later that day, Grimaldi had to stand in a ditch full of what he assumed was water to reach a row of plants with the sprayer. That night, when he went to wash his feet, his toenails fell off, just like flakes of soap. Grimaldi saw that when workers complained, they lost their jobs. The primary lesson on how to handle pesticides was not to utter a word about them.

Although the short-term symptoms of pesticide poisoning often pass quickly, virtually no hard scientific research has been done on the long-term effects. One of the difficulties in observing these effects is that fruit and vegetable pickers are migratory. The turnover in Florida can be as high as 40 percent every year as Hispanic workers leave to go home after earning enough money or simply give up trying to make a go of it in the United States. But plenty of surveys and anecdotal observations raise red flags about the health problems that await today's workers in the decades to come.

Hispanics are relatively recent arrivals on the migrant scene. Decades ago, local African Americans usually performed the same low-paying, excruciating work. In some areas, they stayed after the agricultural jobs moved away. Many are now retired. One such community can be found near Lake Apopka, just north of Orlando. Some farmworker advocates are convinced that the chemicals applied to the produce that these workers picked decades ago are responsible for the litany of medical horrors the community struggles with today.

Leaning on her cane in the scorching midday June sun, Linda Lee matter-of-factly listed her medical conditions: diabetes, lupus, high blood pressure, emphysema, and arthritis. Her hip had to be replaced and her gall bladder removed. Her kidneys failed, so she had a transplant. She also had two corneal implants. Asked what caused her woes, the fifty-seven-year-old African American resident of Apopka didn't hesitate: For nearly a decade as a farm laborer on the shores of Lake Apopka in the 1970s and 1980s, she was routinely exposed to agricultural chemicals as she worked in the fields. "Plenty of my old friends and neighbors got what I got, and a lot of them got stuff I don't want to get," she told me.

In a survey of workers conducted in 2006, eight years after the Apopka farms were closed for good, the Farmworker Association of Florida found that 92 percent of the roughly twenty-five hundred African Americans, Haitians, and Mexicans with whom Lee toiled had been exposed to pesticides through a combination of aerial spraying, wind drifting from applications on adjacent fields, touching plants still wet with pesticides, and inhaling pesticides. Fully 83 percent of those queried reported that their health was only "fair" or "poor." They complained about arthritis, throat problems, diabetes, persistent coughing, recurring rashes, miscarriages, birth defects, and childhood developmental difficulties—all conditions that research studies have linked to the agricultural chemicals that were applied in the area. In a state where the average incidence of birth defects is 3 percent, 13 percent of the Apopka workers had a child born with a defect.

I came to the shores of Lake Apopka at the insistence of Jeannie Economos, the pesticide safety and environmental health project coordinator at the Farmworker Association of Florida. She is in her late fifties but still wears her mane of curly hair in the free-flowing fashion of the hippie era, and she still holds the innocent belief that caring people can make constructive changes to the system. Economos, who signs her e-mails "Yours in solidarity," wanted to give me the "pesticide tour," a firsthand look at one of the country's most extreme examples of what happens when agribusiness shows utter disregard for the environment and workers.

Located fifteen miles northwest of Orlando, almost in the shadow of Disney's Magic Kingdom, Lake Apopka has had many claims to fame. Roughly circular and measuring about ten miles in diameter, it is the state's fourth largest lake. For a time in the first half of the twentieth century, it was nationally famous for its trophy largemouth bass, and twenty-one lodges sprang up on its shores to cater to anglers from around the world. But by the 1980s, Apopka had earned yet another distinction: It was the Sunshine State's most polluted large lake. By then the fabled bass were extinct. Blame for the declining water quality was not hard to assign. In 1941, as part of the wartime effort to produce more fruits and vegetables, nineteen thousand acres of swamp on the lake's north shore were drained to make way for "muck farms" in the rich soil. During the growing season, farmers pumped water in and out of the lake depending on irrigation requirements and rainfall amounts. In the off season, they allowed the lake to flood the fields to replenish the soil and prevent wind erosion and weed growth. With each cycle, the water not only picked up the chemical fertilizers but also the pesticides. Nourished by nitrogen and phosphorus from the fields, the water in the lake turned pea green. The only surviving fish were tough, trashy, minnowlike gizzard shad.

By 1996, the situation had become so dire that the Florida government bought out the big landowners and closed down the farms. The fourteen landowners were paid $103 million for property and equipment. (In one sweet deal, a farmer sold the government a vegetable cooler for $1.4 million and then bought it back at auction for

$35,000.) The twenty-five hundred workers, who often had families that lived with them on the land, got nothing other than the order to clear out. They were not retrained for new jobs because the powerful farmers feared that educated workers would abandon the fields before the last carrot or tomato was picked.

In the winter of 1998, the St. John's River Water Management District decided to reverse the usual pattern of water flow and flood the recently acquired land in the winter to attract waterfowl. Sure enough, that year the Audubon Society tallied the largest Christmas count of migrating birds ever recorded for an inland location. The joy was short-lived. By the end of the winter, more than one thousand fish-eating birds had died—blue herons, white pelicans, bald eagles. It was one of the worst bird-death disasters in U.S. history. A $1.5-million scientific investigation was launched. After a few years, researchers determined that the cause of the deaths was pesticide poisoning. Investigations into the health of alligators also revealed disturbing signs. Males had stunted penises and high levels of estrogen; females had high levels of testosterone. Reproductive rates were far below normal. Again, pesticide poisoning was the cause. But while the research money flowed into looking into the causes of reptile and bird illnesses, not a nickel was spent on examining the laborers who spent their lives working, eating, and sleeping on the contaminated land.

"It's painful trying to get up in the morning and get from one day to the next," said Lee, as we walked along a sandy track through the now-overgrown fields. Even though a dozen seasons had come and gone since the last pesticide-spraying tractor, signs read, "Warning. Visitors must stay on roads. No fishing allowed on this property. These lands were former agricultural land that were subject to regular use of agricultural chemicals, some of which, such as DDT, are persistent in the environment and may present a risk to human health." Lee received no such warnings when she went into the fields to pick corn, cabbages, carrots, greens, and tomatoes, receiving twelve cents to pack a box of corn, fifteen cents for a box of greens.

We finally reached a small park by the lakeshore and took shelter in the shade of a live oak, welcome respite from the scorching sun and unbearable humidity. Cicadas trilled from the scrubby brush that has replaced the rows of vegetables. There was no wind. The water, while not pea green, was khaki colored and opaque. It was high noon on a sunny summer day in the middle of a metropolitan area of two million people, and there was not a soul on the entire fifty-square-mile lake.

The effects of pesticides can travel far beyond the boundaries of Florida's tomato fields, reaching people who have never touched a crop. One Sunday morning, the Reverend Gladys Herrera had to stop midway through service and evacuate all ninety members of her congregation at El Calvario Fuente de Vida, a church in Naranja, a town in an agricultural area just south of Miami. "Sister, I feel sick. I feel bad," said one worshipper. Others reported dizziness, tickling in their throats, and itchiness in their ears and eyes. Kids started coughing. Some vomited. Although no one had warned Herrera, a nearby farm had applied methyl bromide to its tomato fields, the same chemical that felled Guadalupe Gonzales. Odorless and colorless in its natural state, it is often mixed with small amounts of tear gas so that it can be detected. Easily dispersed into the air, the fumigant had drifted in through the open windows of the church.

Subsequent air tests near El Calvario conducted by the Farmworker Association, the Florida Consumer Action Network, and Friends of the Earth, an environmental group, showed that levels of the toxic gas drifting off nearby fields had risen as high as 625 parts per billion, three times the maximum allowable amount set by the government of California (Florida has no standards) and more than ten times the minimal risk level set by the Agency for Toxic Substances and Disease Registry, a branch of the U.S. Department of Health and Human Services.

What happened on that Sunday morning in Naranja is far from rare. In 2009 the Florida Department of Agriculture and Consumer

Services initiated thirty-nine investigations in response to allegations of pesticide drift similar to that experienced by Herrera's congregation. Worshippers at a Baptist church in Homestead, not far from Naranja, were exposed to chemical drift. A schoolteacher in Sarasota was forced to take medical leave after pesticides drifted into the building in which she taught, where five hundred elementary students attended classes. Throughout the state, labor camps, recreational vehicle parks, and retirement communities have sprung up adjacent to, or even within, fields where pesticides are routinely sprayed. In 2007, when the U.S. Environmental Protection Agency proposed establishing buffer zones one hundred feet to one-quarter mile around fumigated fields, Florida farmers cried foul. "This will kill agriculture," insisted Fritz Stauffacher, compliance safety director for West Coast Tomato, which was farming four thousand acres in Florida. Another West Coast executive explained that "growers use land right up to the boundaries." Despite the protests, the new rules went into effect in 2008, but Economos and other antipesticide advocates contend that the only sure solution to the drift problem is to ban fumigants outright.

Not only is methyl bromide a potent poison to humans and wildlife, it is also one of the leading causes of the depletion of the atmosphere's ozone layer, the part of the stratosphere that absorbs ultraviolet radiation from the sun, radiation that causes skin cancer. The bromine in methyl bromide is fifty times more destructive to the ozone layer than the chlorine found in chlorofluorocarbons, which have been banned from production since 1996. Under the terms of the Montreal Protocol on Substances That Deplete the Ozone Layer, the use of methyl bromide was also supposed to have been phased out completely in the United States by 2005. But because of a loophole in the treaty, Florida tomato growers have been granted a "critical use exemption" that not only allows them to use stockpiled methyl bromide but even purchase newly manufactured supplies. The U.S. Environmental Protection Agency claims that otherwise it would be financially impossible for farmers in the state to raise tomatoes. Without the tomato growers' favorite fumigant—which, according to a 2009 report by University of Florida researchers, is still applied to 80 percent of Florida's tomatoes—the agency claimed that there would be a 20 to 40 percent drop in yields. Five years after the so-called "ban," millions of pounds of methyl bromide are still injected into Florida's farmland every planting season. Ag-Mart, the company that operated the fields where the mothers of the three deformed Immokalee babies worked, voluntarily stopped using five of six chemicals that had been connected to birth defects in animal experiments. But it continued to use one of those mutagens because there was no cost-effective replacement. That chemical was methyl bromide.

Antipesticide advocates claim that the exemption is unnecessary. They note that there are several alternatives to the ozone-destroying fumigant, both chemical and nonchemical, and they dispute the claims that not applying it would result in catastrophic crop loss. In one study on processing tomatoes, a team led by Karen Klonsky of the University of California, Davis found that organic production did provide significantly lower yields, but it also cost growers less money to get those yields. When everything was netted out, the profit from organic tomatoes was only 10 percent less than the profit made on the chemically treated ones. More to the point, the study showed that both methods could be profitable.

In Florida, nematodes are one of the main excuses given for application of methyl bromide. But the University of Florida's *Vegetable Production Handbook for Florida 2010–2011* describes several nonchemical alternatives. These practices include rotating tomatoes with crops that are not prone to nematode infestation or simply letting fields lie fallow. Both approaches deprive nematodes of food. Adding animal manure and poultry litter to soil or disking in certain cover crops not only increases fertility but raises levels of ammonia compounds that suppress nematodes while benefiting populations of microbes that compete with the pests without harming crops. Nematodes can also be killed by flooding fields between crops. "Soil solarization" is a promising new technology that involves spreading clear plastic over rows during fallow periods between crops to heat the soil to levels that kill potential pests.

Unfortunately, it appears that Florida growers are showing more interest in an alternative to methyl bromide that many scientists view as one of the most toxic compounds employed in chemical manufacturing—so carcinogenic it has been used to induce cancers in laboratory cell cultures. Called methyl iodide, or iodomethane, the fumigant was approved in 2008 by the George W. Bush–era U.S. Environmental Protection Agency, despite a letter of warning signed by fifty-four of the world's most prominent chemists and physicians, including five Nobel Prize–winning researchers. In their letter, the scientists noted that agents like methyl iodide are "extremely well-known cancer hazards" and that "their high-volatility and water solubility" would "guarantee substantial releases to air, surface waters, and groundwater." Although methyl iodide does not punch gaping holes in the ozone layer, the scientists reminded the agency that its own research had shown methyl iodide to cause "thyroid toxicity, permanent neurological damage, and fetal losses in experimental animals."

Four states have the power to override the U.S. Environmental Protection Agency's approval of a pesticide. Of these, New York and Washington refused to allow application of methyl iodide. California and Florida gave the new fumigant their blessing. A report by the federal government said, "Iodomethane formulated with chloropicrin (one of the Pesticide Action Network's 'Bad Actors' in its own right) has shown good efficacy against key tomato pests . . . in a number of trials." The same report said that methyl iodide would be technically feasible for Florida tomato growers to apply, and the only major impediments were that it costs more than methyl bromide and that some time will be required for Florida growers to make the transition to methyl iodide.

There are already signs that the transition, if it takes place, could have serious health effects for Floridians—and not just those who work in the fields. Tests conducted by Tokyo-based Arysta LifeScience, the maker of methyl iodide (which is sold under the trade name Midas), on wells in the Sarasota area near fields where the fumigant had been applied revealed disturbing results. Iodide, the chemical created when methyl iodide breaks down, was found at concentrations anywhere between six and fifteen hundred times the amounts normally found in fresh water. High levels were also found in the air nearby. According to Susan Kegley, a scientist with the Pesticide Action Network, iodide can cause miscarriages, fetal death, and development disabilities in babies. Yet Davis Daiker, a scientific evaluation administrator with the Florida Department of Agriculture and Consumer Services, told the Associated Press that evaluating the meaning of the water tests at that point was "premature."

It seems like an unfortunate choice of wording, in light of Florida's tragic history of agrichemical use and reproductive disorders, most poignantly those that affected the migrant community in Immokalee.

Andrew Yaffa began work on the case of Carlitos Candelario with a huge cloud hanging over the lawsuit. The baby was an American citizen, but Herrera and Candelario were undocumented workers. At any time, immigration authorities could deport them back to Mexico. Without the parents as witnesses, any case Yaffa could cobble together would go nowhere. The opposing lawyers might even try to arrange for deportation. As a precaution, Yaffa decided to keep the location where the family was living a secret.

Yaffa also knew that Ag-Mart, the company that operated the fields in which Herrera had worked, would fight back with all its resources. After all, millions of dollars were at stake. "I knew Ag-Mart's attorneys would do everything they could to say that the child's condition was genetic," he said. Determined to spare no expense in hiring expert witnesses, Yaffa contacted Dr. Aubrey Milunsky, a geneticist and professor of human genetics at Boston University School of Medicine. Milunsky's laboratories are recognized worldwide as a referral center for prenatal genetic diagnosis. Among the more than twenty books the physician has authored or edited is a major reference work entitled *Genetic Disorders and the Fetus: Diagnosis, Prevention and Treatment*. Milunsky agreed to fly down to Miami. He administered a physical examination of Carlitos and took tissue samples from the baby and his parents back to his Boston laboratory, where he drew out DNA samples. Milunsky's conclusion: Genetics could not have caused Carlitos's condition. "Once we ruled out genetics," Yaffa said, "it became an environmental exposure case."

To marshal the evidence for the trial, Yaffa assembled an impressive panel of experts from around the country. Examining the cluster of deformed births in the small community during such a short period of time, Dr. Omar Shafey, an epidemiologist with a PhD from the University of California Berkeley and now a professor of global health at Emory University, told Yaffa that such a "cluster" could not possibly have happened by chance. There had to be some cause. Dr. J. Routt Reigart, now director of general pediatrics at the Medical University of South Carolina and formerly chair of both the American Academy of Pediatrics Committee on Environmental Health and the U.S. Environmental Protection Agency's Children's Health Protection Advisory Committee, stated that in his opinion Herrera was exposed to a "witch's brew" of herbicides during the early stages of her pregnancy. Dr. Kenneth Rudo, a toxicologist with the North Carolina Division of Public Health, said, "It does appear that an association more likely than not occurred between the exposure of these women to these teratogenic pesticides in the Ag-Mart fields and the adverse developmental effects observed."

But even with the weight of expert testimony building up in favor of Candelario and Herrera, a great deal depended on how the couple would comport themselves before a jury in an American courtroom. Until coming to the United States, the young couple had spent all their lives in Huehuetono, a village of a half dozen streets along a twisty, little-traveled road in the mountains of the Mexican state of Guerrero. Having grown up speaking in their Native American Amuzgo language, which is spoken by only forty-five thousand people living in parts of Guerrero and Oaxaca, Herrera and Candelario were barely literate, knew no English, and had only rudimentary Spanish.

On the morning of June 23, 2006, the two opposing legal teams met to take the parents' depositions around the marble-topped conference table at Yaffa's partnership's headquarters in Miami's Coconut Grove neighborhood. The fifteenth-floor penthouse offices offered sweeping views of Biscayne Bay and the northern Keys. Jeffrey Fridkin, of the Naples firm of Grant, Fridkin, Pearson, Athan, and Crown represented Ag-Mart. With over a decade more experience than Yaffa, Fridkin was a formidable opponent. He is listed in the commercial litigation section of *The Best Lawyers in America,* a peer-reviewed publication that is the legal equivalent of the Academy Awards.

Through a translator, Herrera painted a picture of a journey that is common to migrant workers in Florida's tomato fields. She had gotten her job with Ag-Mart when a recruiter—in Herrera's words a *pollero,* or "chicken smuggler" (also known as a "coyote")—came to her village and offered to take her to the company's farms in North Carolina as part of a larger group of workers he had hired.

Despite her unsophisticated background, there were early signs during the four-hour deposition that the young mother would be a credible witness. Showing her a document called "Chasing the Sun Training Acknowledgement," which referred to a pesticide-handling training film that new Ag-Mart employees were supposed to watch before entering the fields, Fridkin asked if Herrera's signature was on the document. She assented.

Fridkin asked, "Do you recall seeing that video?"

"No, no. I did not see any video," Herrera answered. "I just signed the papers that they gave me."

Fridkin moved on to ask her about a questionnaire that Ricardo Davilos, an employee of the Florida Agriculture Department, filled out after the birth of Carlitos. Quoting directly from the document, Fridkin, addressing Herrera in the third person, said, "The next sentence reads, 'I have never been sprayed and was not made to work in a field that was sprayed.' Did she tell that to Mr. Ricardo?"

"No, I didn't say that."

"What did you tell Mr. Ricardo about being sprayed or not being sprayed?"

"Well, I told him that when—every time they would spray, we would be there picking tomatoes, and we'd feel badly. And then we'd get headaches, earaches. Our eyes would burn. And, also, I—I would get sore throats. I always felt like it burned me and my stomach as well, and I would get a rash on my skin."

"Is that what she told Mr. Ricardo?"

"Yes, that's what I told him, but I don't know how he wrote that down."

"But she did tell him about getting rashes on her skin?"

"I told him about it. My husband has it to this day, he had some, like, blotches."

"Okay. So we are clear, everything that she has just said about how she felt, her eyes burning, are all things that she told Mr. Ricardo at the time that she put her signature onto Exhibit Number Six?"

"I did tell them, but they did not tell me what they wrote down because it was all in English."

"So did Mr. Ricardo ever tell her what he was writing down as her statement?"

"No."

"The next sentence reads, 'I notice when you picked tomatoes all day, your clothes would be all green, and you would smell.' Is that something she said to Mr. Ricardo?"

"Yes," Herrera responded.

After a few more questions, Fridkin asked, "Was it ever enough spray to make her clothes wet?"

"Well, yes," Herrera said. "Like I said awhile back, when it's very, very hot. And then it does wet you completely, the body, the face, the hair."

"Was there ever a time when the spray, as opposed to sweat, made her wet?"

"Yes."

"And would that be true on many occasions?"

"Yes."

"And would that be true when she was picking that she would get enough spray that it would make her wet even with no sweat?"

"That's right. Because when you put your hands in the plants, immediately it sticks to you."

"Does she wear gloves?"

"No, I never use them because we didn't have enough money to buy the gloves."

"And when you were tying and staking, were you sprayed to where you would get your clothes wet with spray?"

"Yes."

"And when you were tying and staking, how many times in a week did you get sprayed to where your clothes were wet with spray?"

"The—the same, two to three times a week."

Toward the end of the deposition, Fridkin asked, "Was this pregnancy planned? Did they plan to have the baby?"

"Yes."

"How did she find out—how did you find out you were pregnant?"

"Well, from the time I arrived, I always felt badly, but I didn't know it was about pesticides. I always had headaches, stomach aches, pain in my eyes, my nose, my throat, my lungs, because I couldn't breathe. But later I felt even worse. So I told my husband that I wanted to have a pregnancy test because I thought that I might be pregnant."

"And how far in the pregnancy were you when you found out you were pregnant?"

"It was about a month. A month and two weeks."

After a break, Fridkin began to depose Candelario, whose testimony painted an even grimmer picture of the conditions under which the expectant couple toiled. "How was it that you would get sprayed to where your clothes were saturated?" asked Fridkin.

"Well, sometimes when they were going—when they're passing close by us on one side of us, and then they'll—they'll spray and then wet us. But it's when they are passing close by."

"And would that happen more than one time a day?"

"Yes. Well, that would happen three times, three times or twice a week."

"Did you ever report that to anybody at Ag-Mart?"

"No. Because the people that are spraying there, they're—they're right there. And sometimes we scream at them to let them know, 'Why are you spraying us? Can't you see that we're here?' So we would run. We didn't have any other option. We would run."

"Would you run two or three times a week out of the field?"

"No, not out. Where—from where they were spraying. So we would run, but the wind would take the spray with it, so we'd run, but it was all the same."

"And that's what happened two or three times a week for every week that you worked for Ag-Mart, whether you were in Florida or in North Carolina; correct?"

"Yes."

The obvious question was why Herrera simply didn't stop working, and her husband gave Fridkin the answer.

"We always had to work the same because we were threatened. Once I was—I told her, 'Stay and rest,' and the person in charge said, 'Are you going to go to work, or are you just going to—just be that way and do nothing.' One day she stayed at home, and then he asked for her. 'Why isn't she here working?' Well, what can I tell him? She stayed behind. And then he said, 'Well, what do you come here for? Do you come to work or not to work? If you're not going to work, then you need to leave the house. You need to leave.'"

By the end of that day of depositions, Yaffa had no worries about how his clients would comport themselves on the witness stand, should the case make it to trial. But there was a weakness in their testimony that an opposing lawyer would surely exploit: Candelario and Herrera had an enormous amount to gain by claiming that they were exposed to pesticides. Yaffa needed a witness who had worked in those tomato fields, who had seen what happened and was willing to talk but had no vested interest in the outcome of the lawsuit. One afternoon his assistant took a call from a woman who identified herself as Yolanda Cisneros. Cisneros told Yaffa that she had worked for Ag-Mart for five years before being summarily fired. She had been in the same fields as Herrero and Candelario, whom she remembered as an extremely good worker. And she had plenty more to say.

Cisneros and I agreed to meet for lunch at a barbecue joint in Immokalee. She was a stocky woman in her mid-fifties, barely five feet tall, but with a forceful, outgoing personality. After a hard handshake, she introduced herself. "I'm the big-mouthed woman you heard about." We took a booth at the back of the room. A jukebox played a string of twangy country tunes as she outlined her life story. Unlike most of the farmworkers in Florida, Cisneros was born on the right side of the Rio Grande, the Texas side. When she was a child, the family moved to Immokalee, and she grew up working in the fields beside her father, mother, and sister as the family followed the harvest each year north to South Carolina and then back to southwest Florida. When Cisneros became old enough to enter school, the family's lifestyle changed. Her father worked in Immokalee for as long as possible each year and then went north alone. "Get an education," he told his daughters. "That way, you'll be able to do some good in the world." The day school ended for the summer, Cisneros's mother packed up Cisneros and her sister and joined her father. The entire family would harvest produce until school started again in the fall.

After dropping out of high school, Cisneros took a series of jobs and eventually found herself employed by a crew boss as the driver of a bus that transported farmworkers from Immokalee to the fields. Within a few years, she had managed to save enough money for a down payment on a used school bus and became the boss of her own crew of ten to forty workers. At the time, she was the only female crew leader in the area. "It was a good old boy network," she told me. The "boys," however, turned out to be not very "good" and didn't take kindly to having a woman among their ranks. Frequently, they hired employees away from Cisneros. Or taunted them: "Oh, you work for a prissy woman's crew." She was given an often-used nickname, "The Bitch." On occasion, she was threatened with physical harm, but she persisted. Farm managers knew that her crew members were good workers. And with crops to be planted or picked, every available laborer was needed, even ones who worked for a woman.

Female laborers, both members of her crew and those working for other leaders, were drawn to Cisneros as a mother figure. She'd take them grocery shopping in the evenings or drive them to doctors' appointments. "They were so far from their homes in a different country. They don't know the language, they don't know the culture," she said. "I knew how the system worked." Women would come to her for advice on what to do when they were being sexually harassed, a common occurrence in the tomato fields. Those experiencing "female problems" were glad to have a woman to talk to about medical issues. If they became pregnant, they would seek her council on whether they should continue harvesting, which involved lugging large plastic flats of grape tomatoes over uneven rows and sometimes required wading across flooded drainage ditches. She often advised them to stay home. But most had no choice. Farm managers would order them back to work.

When I asked Cisneros if she had ever seen anybody sprayed, she answered without hesitation. "All the time. It's part of life out there," she said. "I would tell the women, when you come home, don't hug your kids. I know you want to, but don't hug your kids until you've at least changed your shirt. Otherwise, if you hug your baby, you're going to rub this stuff on it. Sometimes they'd come out of a row of tomatoes and their clothes would be soaked. I'd ask them, 'Why are you sweating so much? Is it that hot?' And they'd say, 'No, the plants were wet.'"

When Cisneros complained to the managers, they would tell her not to worry. She accepted that, until one morning when her crew was covering soon-to-be-planted rows with plastic. Right in front of them, one of the managers, "an Anglo," was driving a tractor that injected soil with a chemical. Rolling out plastic over the rows a few feet behind the tractor, she and her crew were told not to worry, even though they would occasionally cough or experience dizziness. The driver stopped and got down to replace an empty pesticide drum, and while he was attaching a new one, a hose sprang a leak. A squirt of white liquid splashed on his leg—a minor occurrence to pickers. But the supervisor screamed as if scalded and immediately tore off his clothes. "I mean pulling off his pants—not even unzipping them," said Cisneros. "People were telling him that there was a lady out there, but he didn't care. He kept screaming and dove into the ditch and started rubbing water all over himself."

Cisneros looked up and saw one of the company's white Toyota pickup trucks racing toward them. Although roads ran around the field's periphery, the driver roared straight across the newly plastic-covered raised rows, his truck bouncing into the air. He jumped from the truck, helped his associate out of the water, and laid him in back of the truck before speeding off in the direction of town. "And here we were all being told, 'It's okay. If you feel dizzy, get some air, walk around a bit.' This man knew what he was spraying, and this was his reaction. When I asked what had happened to him, they just told me that it was nothing. He's just gotten a little scared. I said, 'Excuse me, this man tore off his clothes, and he's sitting in his underwear, and he has red marks all over his legs—um, something's wrong here.'"

As crew boss, Cisneros's duties were mostly managerial. She received daily instructions on what field to report to with her workers and what the day's work would be—picking, tying, pruning. She directed her team to the designated rows, then briefly demonstrated what she wanted done and how they should do it. On days when they were harvesting large slicing tomatoes, she stationed herself on top of the opened-back truck into which workers dumped baskets of fruits. Her role was quality control, making sure that the tomatoes were of the right size and at the appropriate stage of maturity. It was a position that normally left her at some remove from the pesticide laden plants. But on one occasion, she felt firsthand what the pickers were experiencing. A sprayer was working in an adjacent field, and as it passed, a gust of wind wafted the chemical mist over to Cisneros. "It was just like somebody had taken a big old can of Raid and looked at me and sprayed it right in my face full blast and never stopped until it got empty," she said, making a gun out of her thumb and index finger and wagging it inches away from my eyes. "It really scared me because they knew something we didn't know, and they didn't want us to know."

After that, Cisneros became more "mouthy," frequently complaining to field managers. When some of the men on her crew reported stomach aches and headaches, the supervisor accused them of getting drunk the night before and having hangovers and ordered them back to work. Other complainers were told to take a short walk for a few minutes to clear their heads, or to go have a drink of water. On another occasion Cisneros saw a sprayer approaching along a row where her people were picking and informed the manager. "He said, 'Fine, I'll have the tractor move to the next row.' I said, 'Excuse me, but we're supposed to go over to that row as soon as we're done here.'" One morning she simply refused to allow a worker who was far along in her pregnancy to board the bus, knowing that the day's duties involved planting seedlings. "One, she couldn't bend over," Cisneros said. "Two, her hands were going to be in those chemicals, and I no longer believed that it was safe." The managers reprimanded Cisneros and ordered her to go back and get the woman and bring her to the field. "I felt really bad for those people out there trying to make a living. They weren't bothering anybody."

By the time the *Palm Beach Post* broke the story of the three deformed babies born to Ag-Mart workers in Immokalee, Cisneros had earned a name as a troublemaker. One day after work, the phone rang. A man from the human resources department at the Ag-Mart head office in Plant City, near Tampa, informed her that her services would no longer be needed. It was the height of harvest season.

The restaurant where we were having lunch had emptied, and on the jukebox Tammy Wynette was getting her second D-I-V-O-R-C-E in a little over an hour. After losing her job at Ag-Mart, Cisneros was offered bus-driving gigs by several crew bosses, but she had lost interest in agricultural work. Eventually she sold her bus and other farm equipment, recouping just enough to pay off her debts. At the time we met, she was unemployed. One of her three grown daughters had been diagnosed with cancer, and helping out with doctors' bills had further strained her budget. She had fallen behind on rental payments, and her landlord knew that Cisneros was facing more medical expenses. He evicted them.

Outside the restaurant, she turned to me. "I agreed to be a witness because I wanted to help that poor little boy. I thought that if Andy could get them money, then at least Carlitos could have a comfortable life."

Donald Long had only recently been appointed president of Ag-Mart when reports about the births of the deformed babies to women who had worked for his company began circulating. Long's career path was similar to that of many executives in the tomato business. Shortly after graduating from the University of Florida with a BS degree in vegetable crops production in 1976, Long began working for a grower in the southwestern part of the state, and stayed at that company for twenty years. When Ag-Mart decided to begin producing tomatoes in addition to its traditional strawberry crop in the mid-1990s, Long was hired. His first title was simply "farmer." Later he rose to vice-president of production and ultimately was promoted to president. When word reached him about the birth defects, he drove to Immokalee and met with the parents at Our Lady of Guadalupe Catholic Church. He offered to help find the fathers work in the area during the off season so that the families would not have to move or split up. He told them he would assist them with immigration issues. He told them that if there was anything that they needed, whatever it was, to let him know. He expressed his regrets for the situation they were in. Subsequently, he ordered that Ag-Mart stop using five chemicals that had been linked to reproductive problems in laboratory tests: metribuzin, methamidophos, mancozeb, oxamyl, and avermectin. What he did not do was admit that his company was in any way responsible for causing the children to be born the way they were.

Yaffa's five-hour deposition of Long started out on a casual note. Where do you live? What do you do for work? How long have you been there? Where'd you grow up? What college did you go to? Yaffa assured Long that he would be given every opportunity to explain himself and that Yaffa was not there to cut him off or to trick him. "I just want to find out what you know," Yaffa said. But the exchange soon became adversarial and, on occasion, argumentative, with Long's lawyer issuing objections and forbidding his client to respond to Yaffa's queries.

After establishing that Long had been a licensed pesticide applier before he left the fields for a desk in the executive offices, Yaffa zeroed in on Ag-Mart's continued use of the pesticide methyl bromide. "Is that a product that if you would expose workers to it they might be at risk?" Yaffa asked.

"Risk of what?"

"Harm, long-term physical effects?"

"I don't know about long-term physical effects. They might be—if they were exposed to it, there might be some instant harm or something involved with that, but long term risks, I'm not—I'm not an expert on what the long-term risk would be."

"Pesticide-related illnesses, workers exposed to methyl bromide are certainly at risk for pesticide-related illnesses?"

"I'm not aware exactly what that risk would be."

"You've used methyl bromide over the course of your career?"

"Yes."

"You and your company continue to use it today?"

"Yes."

"It's a deadly product?"

"Correct."

"It's a restricted-use pesticide?"

"Correct."

"You need to have a license to purchase it?"

"That is correct."

"There are certain limitations and restrictions on how you use it?"

"Correct."

"The reason there are such limitations and restrictions is why?"

"So that it is not in the hands of someone that has no training of how to use that product."

"And?"

"And that—I don't know—I mean, it's a restricted-use pesticide which means it could have some harmful effect on a person."

"It carries a warning with a skull and crossbones, doesn't it?"

"That's correct."

"It is a Class I pesticide. Isn't that right?"

"I'm not sure about that."

"Do you know what a Class I pesticide is?"

"It's a very harmful pesticide."

"One that is likely to cause death or serious injury if one is exposed?"

"Correct."

"And your company, over the course of doing business, has and continues to use Class I pesticides in its business, correct?"

"Correct."

"Each and every one of those Class I pesticides carries with it the warning of likely death or serious bodily injury if exposure occurs?"

"Correct."

"And for that reason, these workers who are going to be working with it and around it need protection?"

"Correct."

"If in fact the protections, as set forth in the Worker Protection Standard, are not followed, all these workers are at risk, correct?"

"If all the procedures are not followed, yes."

"Each and every one is mandatory, correct?"

"Correct."

"And if each and every one is not followed, your workers are at risk of exposure?"

"Not necessarily to a risk of exposure."

"What do you mean? Do you think it's okay to cut corners?"

"No."

"Do you think you're obligated to follow each and every mandate as set forth in the Worker Protection Standard?"

"Without a doubt."

"All right. Do you agree that it would be wrong to violate and cut corners?"

"Yes. It would be wrong to violate and cut corners."

"You agree that to cut corners would be in violation of both federal and state law?"

"That's correct."

"You would agree that to cut corners would put your employees at risk. And when I say 'at risk' I mean for serious bodily injury or harm from the pesticides?"

"Um, let me see how to answer this. If—I don't know how to answer this. I think that anything that would expose the worker directly to the pesticide illegally, or by cutting a corner—which I don't believe—we did not do, and we do not have the practice of doing that—would put someone at risk, if we were to cut that corner and put someone at risk."

"And when you make that statement you'll agree that if in fact corners are being cut and the Worker Protection Standard is not being followed, these employees are at risk for developing the long-term sequels and effects of these pesticides?"

"I don't know whether they are developing long-term effects. I'm not a scientist. I don't know that."

"You and I can get beyond that right here at the outset, okay? You're not a scientist, but you know that there is a question about the long-term effects of pesticide exposure to everybody that works with them in and out—correct?"

"I don't know that there is a long-term effect to pesticide exposure."

"Have you done any research on this topic?"

"Have I?"

"Yeah, you. Mr. Long. Don Long."

"Yes, I have looked at what is out there."

"And have you looked to see whether there are animal studies which relate to long-term exposure and birth defects?"

"But they haven't related—they have for animal studies."

"Are you telling me that you think they are going to do that kind of study for humans?"

"No, no, I don't think they are going to do that kind of study for humans."

"So in regards to the pesticides that you use day in and day out, as you sit here today you are aware that there are, in fact, studies linking animals who are exposed to these pesticides to birth defects?"

"Yes, there are studies."

"This isn't new to you?"

"No, no, this is not new."

"You've known about this for a long time?"

"Yes."

"Years?"

"Correct."

"Long before these babies were born to your employees with birth defects?"

"That's correct."

Having established that Long knew that studies existed linking pesticides to birth defects in animals, Yaffa turned to the company's policy about allowing pregnant women to work in its fields. "Tell me about restrictions, limitations, rules, if any, that Ag-Mart has to keep pregnant women from working with methyl bromide when it's being injected into the soil."

"We don't have any rules to keep pregnant women from working with that."

"Why not?"

"We do not have any rules to keep anybody from working with it there. We do not discriminate against people for being pregnant."

"Do you call that discrimination or protection?"

Long went on to make the point that employees were aware that risks were involved in agricultural work and took those risks into consideration before coming to work.

"Do you tell these people what is being sprayed when?" asked Yaffa.

"Yes. There is a list of what's—what applications have been out there. There's a central posting for that."

"How many of these people don't read?"

"Probably quite a few of them."

"Okay, so in regard to your central posting that means nothing to those people, correct?"

"Probably not."

Another heated exchange erupted when Yaffa questioned the reasons and the timing for Ag-Mart's withdrawal of the five pesticides after the Immokalee deformities became public.

"Why did it take that for Ag-Mart to be proactive and take that step?"

"Because at Ag-Mart, Santa Sweets, that—I think that the products that we were involved with were consumer-based products, and I think that there was a misconception between the consumer and people that maybe believed that if they ate the product it might create a birth defect."

"So I want to make sure I get this crystal clear. The change in pesticide policy was made out of concern that there was a misconception on the consumer's side?"

"Yeah, and that if we needed to create a—a safe working environment for employees is to whatever changes we needed to do that."

"If, in fact, there was a misconception on the consumer end that would directly affect Ag-Mart's profits?"

"It could affect our business, yes."

"So Ag-Mart's concern about their profits came directly into play in their decision to stop using pesticides known to be linked to cancer in lab animals?"

"No, and the perception of the business within—within the ad community and the worker community, it was a whole round situation of what was—what was going to be better for our workers and what was going to be better for our product."

"Yes, but profit came directly—"

"No."

"Let me get the whole question out. You knew for years that these pesticides were linked to birth defects in lab animals. We talked about that . . . Knowing the risk was there, why not be proactive and take that step before you have three women bearing children with such horrific defects?"

"Well, the three women were not all—I don't believe that –this is my belief, so I—I—don't believe that the pesticides caused the birth defects. I believe that the pesticides have been tested to cause birth defects in animals, but I don't believe pesticides caused birth defects in those three women."

To Yaffa's disappointment, Long's belief was given official legitimacy ten months after Carlitos's birth when the Collier County Health Department issued the results of its investigation into the Immokalee deformities. After consulting with a leading Florida geneticist, department officials said that pesticides were unlikely to have been the cause of the cluster of birth defects. "We were unable to make the link between the pesticide usage and the birth defects in these particular women," Joan Colfer, the director of the department, told the *Naples News.* "That doesn't totally rule it out. It's just that we were unable to make the link." Which was the very link that Yaffa, who contended that the Collier investigation amounted to nothing more than a non-investigation, needed to make in the minds of a jury.

A critical break in the suit came when Yaffa found out that Maria Meza, the woman who had given birth to Violetta, the Tower Cabin baby who died a few days after being born, had aborted an earlier pregnancy at the suggestion of doctors. That pregnancy had also begun while she was working in the Immokalee fields. Like Carlitos, the aborted fetus had neither arms nor legs. The odds of finding two cases of tetra-amelia in a small community were extremely remote. Then an investigator who was interviewing former Florida farmworkers in Mexico reported that he had found a woman in an isolated village who had been working in Immokalee several years earlier and had given birth to a stillborn child. That baby was also limbless. Although the statute of limitations had passed and she had nothing to gain, the woman agreed to testify. A member of Yaffa's legal team traveled to the village and deposed the woman. It was the legal equivalent of lightning striking in the same place three times. A trial date in the case of *Francisca Herrera and Abraham Candelario as parents and natural guardians of Carlos Herrera-Candelario v. Ag-Mart Produce, Inc.* was set.

On Friday, March 21, 2008, nearly three years after Yaffa took on the case, he and Ag-Mart's legal team met to try to reach a settlement. Negotiations lasted into the night, and when the two parties emerged, they had a deal in hand. Ag-Mart admitted no guilt but agreed to pay a substantial sum. Both sides agreed that the exact terms of the settlement would remain sealed. "The amount assures that Carlitos will have all the care he needs for the rest of his life," Yaffa said. At the request of Candelario and Herrera, the money was placed in a life-care plan, overseen by a trustee who is charged with making sure that any funds that are spent directly benefit the boy. The family was able to purchase a small bungalow in Immokalee. Carlitos moves about in a custom-designed wheelchair and is enrolled in school. He is an intelligent, out-going little boy who is well liked by his classmates. Candelario still works in the fields, but Herrera was able to stop picking tomatoes and devote all her time to her boy. Two years after the settlement, she discovered that she was pregnant with a second child. Carlitos's little sister was born in 2010, "a beautiful baby" according to Yaffa. "Carlitos's birth stands for a whole lot more than a child born without arms and legs," Yaffa said. "This child has changed the system."

Or part of the system. Sadly, pesticide exposure is only one of a long list of abuses that the men and women who pick our winter tomatoes have to endure.

FROM THE HANDS OF A SLAVE

Should you want to experience culture shock in one of its starkest forms, take the drive from Naples, Florida, to Immokalee. Your journey will begin in a city of handsome boulevards lined with stately palms and bordered by well-trimmed street-side planting strips of tropical shrubs. Visitors and winter residents alike dine at outdoor cafes and shop at expensive boutiques, antique stores, art galleries, and high-end chains like Cartier, De Beers, Saks Fifth Avenue, Gucci, Tiffany, and Hermès. Hundreds of yachts create traffic-jam conditions in the blue waters of the harbor, and squadrons of private jets whistle in and out of the municipal airport. "Ultra high-end luxury homes" in town listed for as high as $24.9 million even after the Florida real estate crash. Beachfront condos can fetch $14 million. In 2008 Moody's rated greater Naples as the country's wealthiest metropolitan area, with an average net worth of $1.7 million.

As you head east on State Road 846, the gated golf course communities become sparser and less grand. The highway dwindles from six lanes, to four, to two. The mansions and shopping plazas give way to more humble developments of bungalows and plain strip malls, and eventually to cypress swamps and dry stands of pine, cabbage palm, and scrubby palmetto. Occasionally, you pass a clearing occupied by a low ranch-style home, its property lines demarcated by a rusty chain-link fence and a padlocked metal gate. Yards are cluttered with faded sport utility vehicles, fat-wheeled pickup trucks with tinted windows, and outboard boats that look like they have been sitting immobile on their trailers for several seasons. Farther inland, human habitations disappear, and the occasional citrus grove abuts the road, its neatly rounded, deep-green trees marching off in straight ranks. What you don't see is tomato fields. But they are there, hidden behind ten-feet-tall berms covered in scruffy vegetation and broken sporadically by access roads festooned with "No Trespassing" signs and guarded by private security men.

Less than an hour after leaving Naples, you round a long curve and enter the city of Immokalee. A few years ago, county officials attempted to bring a veneer of vaguely Latino urbanity to the main drag by laying down a paving-stone median and crosswalks, planting some small palms, and erecting fake antique streetlights. Maybe the hope was that tourists passing through would not notice conditions a block or so away. Downtown Immokalee is a warren of potholed lanes leading past boarded-up bars and abandoned bodegas, moldering trailers, and sagging, decrepit shacks. The area is populated mostly by Hispanic men, although you will see the occasional Haitian woman (a holdover of an earlier wave of ethnic farm laborers) walking along the sandy paths that pass for sidewalks with a loaded basket of groceries balanced on her head. Scrawny chickens peck in the sandy yards, and packs of mongrels patrol the gaps between dwellings, sniffing at the contents of overturned garbage cans. Vultures squabble over a run-over cat lying in the middle of a street. Immokalee's per capita income is only $9,700 a year, about one-quarter of the national average. Half of the people in the city of fifteen thousand live below the federal poverty line. Two-thirds of the children who enter kindergarten drop out of school without high school diplomas.

Your chances of becoming a victim of violent crime in Immokalee are six times greater than they are in the average American municipality. On the crime index, where zero is the rating given to the most dangerous areas in the United States and one hundred is the rating given to the safest, Immokalee comes in at one. Even the police there are sometimes criminals. Glendell Edison, a deputy sheriff who patrolled Immokalee for fifteen years, was sentenced to ten years in prison after being convicted for extorting money from drug pushers and possessing cocaine and crack. Florida's largest farmworker community, Immokalee is the town that tomatoes built.

As a United States attorney for Florida's Middle District based in Fort Myers, Douglas Molloy has had more than a decade of experience dealing with crime in Immokalee. More specifically, Molloy, who is in his early fifties and has wavy salt-and-pepper hair and a deeply lined face, has gained an international reputation by specializing in prosecuting an act that was supposed to have vanished from the United States 145 years ago. At any given time, Molloy works on six to twelve slavery cases. Immokalee, he says, is "ground zero for modern-day slavery." He also says that any American who has eaten a winter tomato, either purchased at a supermarket or on top of a fast food salad, has eaten a fruit picked by the hand of a slave. "That's not an assumption," he told me. "That is a fact."

The one-story, L-shaped house at 209 South Seventh Street stands in stark contrast to the couple of dozen trailers that surround it on three sides. A handsome royal palm shades the front lawn. The dwelling is fairly new, well painted, and in far better repair than the average Immokalee residence. From 2005 to 2007, Lucas Mariano Domingo lived at that address. New to town, broke, and homeless, he faced the prospects of many recently arrived migrants—sleeping at missions and in encampments in the woods and sustaining himself through once-a-day trips to the local soup kitchen until he amassed enough money to get a room in one of the trailers and buy his own food. Domingo must have thought it was a great stroke of luck when

Cesar Navarrete, a strapping twenty-four-year-old Mexican he met on the streets of Immokalee, not only gave him a job but invited him to crash on his family's property on South Seventh and even offered to front him some pocket cash. For fifty dollars a week, Navarrete's mother, who also lived in the house, would provide meals. Domingo could pay her after his first check—a handsome sum. Navarrete, who ran a harvesting crew with his brothers, was willing to pay Domingo one dollar for every bushel-size bucket of tomatoes he picked, more than twice what many crew bosses were offering at the time. As for Domingo's lack of documentation, no problem. Navarrete knew someone who could get him false papers. Domingo, a Guatemalan in his thirties, had come to the United States with the dream of making enough money so that he could send some home to care for a sick parent. With a little quick calculation, he determined that he'd be clearing $200 a week from Navarrete, leaving him with plenty of spare cash to wire back to Guatemala.

From the outset, it became apparent that Navarrete's promises were too good to be true. Domingo's twenty-dollar-a-week rent wasn't for a room with the family in the neat house but for shared space with three other workers in the back of a box truck out in the junk-strewn yard. It had neither running water nor a toilet, so Domingo and his "room" mates had to urinate and defecate in one corner. It turned out that there were about a dozen other men living behind the Navarrete residence, some in trucks like the one Domingo now called home, others in old vans, and yet others in a crude shack. Navarrete's mother's promise to provide food turned out to be two meager meals a day—eggs, beans, tortillas, rice, and rarely some sort of meat—only six days a week. "The food was terrible," said Jose Hilario Medel, a two-year veteran of the Navarrete crew whom I was able to interview. He made a gagging sound. "Some days you'd maybe get four tortillas—nothing else. Often the food would run out before everyone got his share."

But Navarrete was generous in one way: He was always eager to extend loans for his crew to buy all the beer, wine, and liquor they wanted, no worries. And pretty soon Domingo, like other members of the crew, found that he had become addicted to the alcohol that flowed so freely at 209 South Seventh. Everything, it seemed, had a price that Navarrete jotted down in a notebook, even activities related to basic hygiene. At the end of hot days of fieldwork, Domingo came home covered in perspiration and pesticides and had to pay five dollars to stand naked in the yard and spray himself off with cold water from a garden hose. His debts soon reached $300.

Still, he was making a dollar a bucket, and by his calculations, after nearly a month of ten-and-a-half-hour shifts, six days a week, he had picked many times three hundred buckets. But when Domingo, skinny and less than five and a half feet tall, brought the subject up, Navarrete said it didn't matter. Domingo was still in debt as far as Navarrete was concerned, and if he tried to leave, he would be caught and soundly beaten. Any crew leader who dared to hire him would get the same treatment. Every week, Navarrete made Domingo hand back his paycheck. After deducting a hefty check-cashing fee and subtracting for rent, food, showers, bottled water, and liquor, he'd hand Domingo arbitrary amounts, twenty dollars one week, fifty dollars the next. Taking a day off was not an option. If Domingo or any of the others in the crew became ill or too exhausted to go to the fields, they were kicked in the head, beaten with fists, slashed with knives or broken bottles, and shoved into trucks to be hauled to the worksites. Some were manacled in chains. According to Medel, one day a crew member couldn't take it anymore and ran away from a field. One of the Navarretes got in his truck to chase him down. When the truck returned, Medel said that the man's face was so bloody and swollen that he was unrecognizable. He could not walk. "This is what happens when you try to get away," the boss said.

The spare legal language of the indictment that eventually came down against Navarrete and five other members of his extended family reveals the brutal conditions Domingo and the other enslaved laborers endured over a period of nearly three years:

- On or about May 2005, defendant Cesar Navarrete stopped paying his workers then told LMD and other workers that if they went to work for anyone else he would find them, beat them and their new employer.
- On an unknown date at the Navarrete property, defendant Cesar Navarrete grabbed AL and locked him in the back of a yellow truck for approximately four hours because AL wanted to leave his employment.
- On an unknown date approximately a month after AL was locked in the truck, defendant Cesar Navarrete hit AL because he wanted to leave his employment.
- In or about 2006 in South Carolina, defendant Cesar Navarrete slapped and kicked LSL because he did not want to work and then locked him in a truck for six hours.
- In or about June 2007, in DeSoto County, Florida, on a morning when LMD did not want to work, defendants Cesar Navarrete and Geovanni Navarrete picked up LMD and threw him in the back of a truck, beat LMD in the body, head, and mouth causing him injury and pain.
- In or about July or August 2007 in Walterboro, South Carolina, defendant Geovanni Navarrete and others known to the grand jury chained PSG's feet together and to a pole to prevent him from leaving their employment.
- In or about September 2007 in North Carolina, defendant Geovanni Navarrete struck APS near his eye causing serious injury to his eye.
- Between on or about June 11, 2007, through November 2007, defendant Cesar Navarrete beat LSL almost daily.
- In or about September 2007, in North Carolina, defendant Cesar Navarrete kicked JVD in the ribs to make him go to work.
- In or about September 2007 in North or South Carolina defendant Cesar Navarrete ordered the beating of DC by others known to the grand jury because he was drunk and did not want to work.

- In or about mid to late November 2007, in Collier County, Florida, defendant Geovanni Navarrete and others dragged ATR out of the shack on the Navarrete property where he was sleeping at night time and severely beat him causing injury to his body and head.
- In or about November 2007, defendant Cesar Navarrete and others known to the grand jury, hit and kicked RRC because he left the Navarrete property without permission and went to the Coalition of Immokalee Workers office.
- On or about November 19, 2007, in Collier County, Florida, defendants Cesar Navarrete and Geovanni Navarrete, and others known to the grand jury beat LMD then locked him, JHM, JVD, and two other workers in the back of a box truck to prevent them from leaving and to ensure they would be available for work the following morning.

That last incident proved to be the beginning of the end for the Navarrete clan. Lying in the back of that truck, unable to sleep during the night due to a severe contusion on his forehead, Domingo noticed a shaft of light glimmering through a gap between the roof and the sides of the truck just as dawn broke. Jumping up and down until he was able to grab hold of the edge and punch his way through, he struggled out onto the truck's roof. Another confined worker, Jose Vasquez, followed him, and together they crept down and found a ladder to help the others escape. For the first time in two and a half years, Domingo was free.

They approached the first person they encountered on the streets, who happened to be a local pusher with several run-ins with the authorities. On the escapees' behalf, he called the police. The first officer to arrive was a member of the drug detail. Fortunately, he had attended training sessions put on by a grassroots workers' group called the Coalition of Immokalee Workers and recognized all the signs of a slavery case in Domingo's story. He alerted his colleague Charlie Frost, who specialized in human trafficking cases for the

Collier County Sheriff's Office. On November 21, 2007, Domingo and his fellow escapees told their harrowing story to the deputy sheriffs, who wasted little time. Armed with search warrants, they raided the Navarrete residence at five o'clock on the morning of November 29, freeing a dozen more workers. On December 5, three of the Navarrete brothers and their mother were indicted for the relatively minor charge of "harboring illegal aliens." Subsequently, they were charged with a "conspiracy to deprive the civil rights of undocumented foreign nationals from Mexico and Guatemala, namely, the right to be free from involuntary servitude as secured to them by the Thirteenth Amendment of the United States Constitution." Or, as Molloy put it in layman's terms, "Slavery, plain and simple."

Human trafficking, or slavery if you prefer, is a very difficult crime to solve. Government statistics suggest that a total of about fifteen thousand new human trafficking incidents take place in the United States each year (no one has a precise figure). Almost the same number of murders occur in the country annually. However, 65 percent of homicides are solved. For human trafficking, the rate is 1 percent. Two things account for this disparity. One is that law enforcement officials are reluctant to charge potential human traffickers unless the case is solid. Acquittal not only exposes witnesses and police sources to possible retaliation, but it sends a message to would-be traffickers that they can get away with this highly profitable crime. Second, in a slavery case involving undocumented workers, there are added hurdles. Without barred windows, whips, and chains, prosecutors often have to base their cases entirely on the testimony of the slaves themselves. In a strange country where they understand neither the language nor the law, workers are reluctant to come forward. In their homelands, cops are often thugs in uniform, so they have good reason to fear police. Without green cards, they face arrest and deportation. Stories abound of slaves making a mad dash for the border moments after being freed. Misguided charitable organizations have even stepped in to help former slaves flee the country before they could give evidence. Traffickers also use threats against victims' families in their home countries to exercise control, according to Detective Frost. In one instance, when a slave ran away, members of his boss's family went to the victim's family in Mexico and informed them that if the escapee did not return to work, "We will kill you. Next time you talk to him, tell him that." After his next call home, the worker returned to his boss. In another example, Frost was ready to proceed with a case when his witnesses began to waver on giving testimony. Worried about the ramifications for friends and family at home, they explained to him, "The traffickers are the law of our village. They have the guns. They make the laws." Finally, many slaves don't come forward because they believe that they are truly in debt. In their tight-knit societies, failure to pay debt is considered to be one of the most dishonorable acts a person can commit. "Slavery is unlike any other crime," Molloy told me when we met in his large corner office in downtown Fort Myers. "Victims don't report themselves. They hide from us in plain sight."

With faux jungles full of parrots and monkeys, performing mermaids, and parks cashing in on every other conceivable theme, you'd think that the last thing Florida needed was yet another "attraction." But in early 2010, the Coalition of Immokalee Workers founded the Florida Modern-Day Slavery Museum. Housed in a box truck almost identical to the one in which Domingo and his fellow crew members were confined, the museum was meant to be a traveling exhibition to take the message to every corner of the state. That effort was so successful the museum embarked on a month-long road trip in the summer of 2010 up the eastern seaboard to Washington, DC, New York, and as far north as Boston.

I had my first tour of the museum one cool, cloudy day in February while it was still a work in progress. As three coalition members wrestled with an iron staircase destined to become the entrance, I hopped up into the cargo area of the truck. At the time it was stacked high with empty cardboard tomato boxes. A plywood sorting table similar to what doubled as a bed for Domingo and his peers ran along one side. Even though a roll-up door at the back of the box was open, sweat began pouring down my face and back and dark stains spread out from my armpits. The heat was stifling. I started to feel light-headed and hopped back outside. If this was what it was like on a comfortably brisk, overcast winter day with the back door wide open, what would it have been like for the workers kept locked in there for hours at a time in the heat of summer?

The coalition was assembling the sort of exhibits that you'd expect to find in a museum dedicated to the history of slavery in this country: chains to manacle disobedient slaves; a coarse-cloth blood-stained shirt worn by a picker who had been pummeled by an overseer in the fields for not working hard enough; wooden-butted pistols that would be drawn as threats, deployed in beatings, and used to shoot escapees. None of these were relics from the pre-emancipation 1800s. They all came from cases the coalition had helped bring to light in the previous decade or so.

Slavery and agriculture have had a close relationship in Florida since European settlers first buried seeds in its sandy soil. But the institution really took root when Britain gained control of the region in 1763, and planters from the Carolina colonies moved into the St. John's River area in the northeastern part of the state to raise crops of rice and indigo. By 1860, just before the start of the Civil War, 44 percent of Florida's 140,000 residents were slaves. When that system abruptly ended in 1865, cooperative local sheriffs obligingly arrested gangs of African American men, typically on bogus vagrancy charges, and rented them out to landowners in "convict lease programs," a good deal for both the municipality collecting the fees and the farmers. "Before the war, we owned the negroes," one planter famously said of the system in the late 1880s. "If a man owned a good nigger, he could afford to take care of him. But these convicts: we don't own 'em. One dies, get another."

After 1923, when Florida and Alabama became the last two states to ban the convict lease system, unscrupulous growers switched over to debt peonage. Workers racked up debts to their bosses through exorbitant charges for rent, food, beer, and cigarettes. When deductions from their wages were made at the end of each month, they found that they had fallen even further behind. Their souls might have been their own, but they owed their bodies to the company store. A landowner of that time told an interviewer for the 1960 CBS documentary *Harvest of Shame*, "We used to own our slaves, now we just rent them." Today, fifty-one years after that film first aired, unscrupulous crew bosses find that time-tested debt-peonage tactics still work just fine.

Situated on the banks of the Caloosahatchee River, about a half hour's drive north of Immokalee, LaBelle is a quaint little town with a handsome brick courthouse and modest, old-style southern homes that slumber in the deep shade of ancient, moss-draped live oaks. For several years in the 1990s, sheriff's deputies working there noticed a disturbing trend. All too frequently, LaBelle's tranquility was shattered by gunfights, some of which brazenly occurred in the main street in front of a bar owned by Miguel Flores, a farmworker crew leader who operated in Florida and South Carolina. Often the combatants were Flores himself and men who had formerly been his lieutenants but had since fallen out. Bodies of unidentified Hispanic males were found floating in the river. Although there was never any evidence connecting Flores to the gunplay and corpses, police began to suspect that Miguel A. Flores Harvesting, Inc., engaged in activities far more sinister than picking tomatoes. But there was no proof, and initial inquiries led to nothing. Officialdom lost interest in Flores.

In the early 1990s, the Coalition of Immokalee Workers was in its infancy. Its leaders were part of the same migratory crews who picked tomatoes and oranges around Immokalee in the winter and traveled north during the warm months. One day in 1992, a tiny woman named Julia Gabriel complained to the coalition members that she had not been paid. The crew she worked on had been ordered to evacuate camp in the middle of the night. The bosses didn't even wait to collect the money owed to them by the farmer who had employed them. The reason for the abrupt departure, Gabriel said, was that the head crew leader had shot an employee who was trying to leave for another job. The gunman's name, according to Gabriel, was Miguel Flores.

The Coalition of Immokalee Workers took the information to law enforcement officials. It required four more years, but in 1996 Flores and his second-in-command, Sebastian Gomez, were indicted on numerous counts, including holding others in involuntary servitude, collecting debts by extortionate means, transporting illegal aliens within the United States, concealing and harboring illegal aliens, and transporting migrant farmworkers in unsafe vehicles. According to court papers, Flores and Gomez had two accomplices, Nolasco Castaneda and Andres Ixcoy, who were stationed near the Arizona/Mexico border. Their duty was to find and hire workers with a particular set of job qualifications. Prospective Flores Harvesting employees had to (1) have no legal documentation, (2) speak no English, and (3) have little or no education. Once recruited, the workers were crammed into vans that had been specially modified to hold up to twenty-six passengers sitting or lying directly on the floor. During the journey from Arizona to worksites in South Carolina and Florida, employees were not allowed to leave the vans. They urinated into plastic jugs.

In the Flores labor camps, the new arrivals were informed that they owed hefty transportation fees for their miserable rides east. Anyone who attempted to leave the camp while still indebted to Flores would be hunted down like an animal and killed. Many modern-day slavers employ an element of macabre theater to intimidate their workers and ensure obedience. Flores was a master of this craft. Reveille each morning at his camps came in the form of volleys of shots fired by his henchmen, aptly called *pistoleros*. The same enforcers would stand constant guard over the workers in the fields. Verbal orders to work harder were punctuated by brandished handguns that were often discharged into the air for added emphasis. If that didn't drive home the message, physical assault did. On one occasion, a worker named Antonio Perez received a severe beating at the hands of Gomez. When his friend, Ramon Pena, attempted to help the badly injured Perez, Flores stepped into the fray, bashing Pena in the head with the butt of a semiautomatic Smith and Wesson pistol and sending him to the hospital for a week with a severe concussion and multiple lacerations to his scalp. Although he was occasionally tossed in jail on minor weapons violations (and promptly bailed out by the big growers for whom he supplied such a large and reliable workforce), Flores ran his operation on the same brutal business model for fifteen years, controlling more than four hundred laborers who worked ten to twelve hours per day and for their efforts cleared about twenty dollars a week.

Ultimately, something good came out of the horrors of the Flores camps. Because of the courage of laborers like Julia Gabriel, who stood up and faced her tormentors in court, and the efforts of the Coalition of Immokalee Workers, the Flores case established a template for a string of prosecutions that would soon follow. Law enforcement began to cooperate with civilian organizations with close ties to the workers, including the coalition, Catholic church groups, and other workers' advocates. As Frost, a wide-faced self-described "white guy" who stands well over six feet tall, put it, "When I walk into town, I stand out like a sore thumb. The coalition members fit right in." The Flores case also spurred Congress to enact the Victims of Trafficking and Violence Protection Act of 2000, which codifies what constitutes slavery in modern-day America and sets forth severe penalties for those who perpetrate it. By the time that law was passed, it was becoming clear that Flores was by no means an aberration in the tomato fields of Florida.

Like Lucas Domingo, Antonio Martinez came to the United States because his parents were both sick. Patchy jobs available to him in Hidalgo, Mexico, provided too little income for Martinez, the only breadwinner in a family of two parents and four younger siblings. So he set out on the same perilous path taken by the vast majority of people picking tomatoes in Florida, the same path taken by seventy-two migrant workers found murdered at a ranch near the United States border in 2010. He paid a *coyote* to be guided northward to a promised construction job in Los Angeles. At the border between Mexico and Arizona, Martinez was passed along to another *coyote*, who led a small band of migrants across the desert for three days. They ran out of food and water after the first day. Eventually, a van met them and drove them to a safe house in Tucson. At this point, the second *coyote*, a man named Chino, demanded more money. Martinez explained that he had none. Chino told him that in that case, the Los Angeles construction job was out, but that he might be able to arrange for tomato-picking work in Florida that would earn him $150 a day, which would enable Martinez to pay what he owed.

To get to Florida, Martinez and sixteen other recent arrivals boarded a van driven by a man whom everyone called *El Chacal*, the Jackal. At the end of the two-day journey, during which the men had only two shared bags of potato chips to eat, El Chacal turned off a highway about fifteen miles short of Immokalee and drove along a secondary road leading into the Corkscrew Swamp, a vast expanse of watery cypress stands interspersed with drier pine and palmetto scrub. The van stopped in front of a decrepit trailer set in an overgrown clearing about a hundred feet back from the road.

El Chacal got out of the van and greeted a man. A heated discussion ensued. Martinez heard El Chacal repeating "$500" over and over again. The stranger shook his head. Finally, Martinez heard him say "$350." A roll of bills was produced, and El Chacal ordered his passengers out of the van and drove off, leaving them at the mercy of Abel Cuello, their new owner. "We were being sold like animals," Martinez recalled.

They were treated like animals, too. At night, the twenty-six men in Cuello's crew were locked inside the two trailers he owned.

Conditions were no better inside than they had appeared from the road. The floor was riddled with holes, providing easy access for cockroaches, rats, lizards, snakes, and other animal denizens of the surrounding swamps. Water drawn from a shallow well outside was rank smelling and foul. The men slept on mattresses on the floor. In the morning, Cuello unlocked the door and drove his crew to the worksites. He never took his eyes off them and threatened to beat them or kill them if they tried to escape. When three of the men made a desperate attempt to flee, he caught up to them and ordered them back to work, screaming, "I lost $5,000 on you. I want my money." Another would-be escapee was pursued and run down by Cuello's vehicle. "I own you," he told the stricken worker. One day about four months after he had begun work for Cuello, Martinez and a few other workers were taken to a convenience store to spend the token sums that they were given. While waiting in the parking lot for the men, Cuello fell asleep, and Martinez escaped. Cuello eventually pleaded guilty and received a lenient thirty-three-month sentence. After his short stint in federal prison, he returned to Immokalee and was immediately rehired as a crew boss at another major tomato company.

Laura Germino is a slender woman in her mid-forties whose family has lived in Florida for six generations. Germino has been with the coalition since its inception in the early 1990s and runs its antislavery program, which received the 2007 Anti-Slavery Award from Anti-Slavery International. In 2010 Germino accepted another award from Secretary of State Hillary Clinton for the group's antislavery efforts. No one at the coalition set out to become an expert at uncovering slavery rings and bringing their activities to the attention of the authorities, Germino told me as she drove north toward the town of Lake Placid, where a slavery gang had operated in the 1990s. But in the course of normal outreach programs, members of the organization were often approached by workers who offered up tidbits of information or sometimes just rumors. "Something's not right about that guy's crew." "A worker got shot up north." Nothing that a police officer would act on, but stories that Germino let percolate in her subconscious, sure that someday something would happen to fill in the blanks. "There's usually a grain of truth to these rumors," she said.

Several relatives of the convicted men remained in the Lake Placid area, and Germino makes occasional trips there just to check things out or maybe spot a boss keeping an unnecessarily keen eye on a busload of workers he has driven to a bodega. She also keeps her eyes open for reasons of self-preservation. In 2010 a particularly violent human trafficker she had helped put behind bars in the late 1990s was released from prison. The same law enforcement officers who used the information gathered by Germino and the coalition to make their case against him neglected to inform her that he had been released and was back on the streets of the town where she lived.

Lake Placid's commercial heart could belong to any wholesome all-American county seat. Neat brick buildings house a library, shops, schools, hair salons, mom-and-pop restaurants, coffee shops, a local museum, a car dealership, and many, many churches. Tourists pull off the highway to spend a couple of hours looking at Lake Placid's renowned historical murals. There are more than forty of them: conquistadors, early Florida cowboys lassoing longhorns, modern-day families laughing together as they blast through saw grass swamps aboard airboats, an early-twentieth-century outdoor prizefight, three happy clowns, graduates, presumably, of Toby's Clown School, a local educational institution. Lake Placid bills itself as "The Town of Murals," or "The Caladium Capital of the World" because 97 percent of the bulbs that produce the colorful plants are grown on twelve hundred acres in the area. Rolling terrain pockmarked with twenty-seven lakes surrounds the community, an idyllic setting for the middle-class retirees who have settled there.

Germino turned the car into a development called Sun 'N Lakes (Sunny Lakes to the farmworkers). Like many inland subdivisions in Florida, Sun 'N Lakes never lived up to its initial promise. Pioneering real estate promoters laid down a grid of streets and began to advertise that inexpensive land was available on easy-to-meet financing terms. Prime lots near a small park and beach on the shores of Lake Grassy were sold, and modest retirement homes sprang up. Less desirable sites farther away from the water remained undeveloped and reverted to scrub. For a half mile or so after Germino and I left the main road, the community of neat two-bedroom homes looked pleasant enough, but soon the houses became shabbier and were replaced first by manufactured homes, then by old trailers haphazardly aligned on their lots. The roads themselves disintegrated from pavement, to pot-holed pavement, to sand. The shoulders provided final resting places for rotting mattresses, old sofas, broken-hulled boats, refrigerators, air conditioners, baby seats, television sets, and pile after pile of bullet-riddled beer cans. Potholes became large puddles that Germino had to ease her car around. Finally, we encountered a pond that spanned the road. "Oops," she said, reversing course until we came to a marginally more promising crossroad. "I think it's this way."

I was relieved when the road grew wider and less bumpy. It was paved again by the time she stopped at a phone booth opposite a convenience store on Highlands Avenue. On the other side of the road, a narrow track led up to five or six trailers. "The store is new," she said, with surprise. "That land was just brush the night they murdered Roblero."

Ariosto Roblero was a Guatemalan who was a van driver for one of several small companies that transport migrant workers from Florida to northern fields, part of a loose, unregulated, but extremely efficient network that with a few vehicle changes can get laborers to New York, Chicago, Houston, Los Angeles, Guadalajara, Mexico City, Oaxaca, and pretty much any other city in North or Central America that has a sizeable Hispanic population. On March 20, 1997, police responding to a 911 call found a man lying face down on the pavement in a pool of blood beside a van. He had been shot once in the back of his head, clearly an execution. Witnesses reported that Roblero, the dead man, was taking a group of workers from the area, where the harvest was wrapping up for the season, to North Carolina. He had stopped at the pay phone to wait for an additional passenger when a car and a pickup truck screeched to a halt. A group of men hauled Roblero out of the van and shot him. Initially, the witnesses told police that they suspected the murder might have been the work of an associate of Ramiro Ramos. A local crew leader, Ramos (called *El Diablo* by his employees) was angry because one of the five hundred pickers in his employ had left the area without having paid off the "debts" he owed. His means of escape was a van similar to Roblero's. Even though Roblero had not been the driver, his murder was intended to send a clear message to anyone who might be foolish enough to provide rides to workers trying to leave Ramos's employment. Unfortunately, the case went nowhere. When the time came for formal interviews with the police, no witness would repeat what he had said during initial questioning.

Germino and I wound our way back to the highway. After about a half-mile, she pulled into the parking lot of a store called El Mercadito and continued her narrative. Almost exactly three years to the day after Roblero had been executed, a similar incident occurred at the store. At a little before midnight, a convoy of four vans that had set out from Immokalee made a prearranged stop at the store to pick up migrants to take them to the next place where crops were ripening. Two pickup trucks barreled into the lot. Six or seven men jumped out. Some held the passengers at gunpoint, while others smashed the vans' windows. Jose Martinez, who owned the vans, was one of the drivers that night. An attacker pistol-whipped him, splitting open his face from the hairline to the bridge of his nose and continued beating him, screaming, "You're the motherfucker who's been taking my people. I'll kill you." Seeing what was happening to his boss, another driver escaped from the fray, called 911, then telephoned his brother, Lucas Benitez, one of the founding members of the coalition. "We are being attacked by men with guns! They look like bosses!" he said. Within the hour, Benitez and Germino arrived.

They realized that the gang leaders' attempts to eliminate vans stopping in areas controlled by the Ramos family was a signal that slavery was involved. But the police weren't eager to pursue an investigation of that complexity. Instead, they rounded up Ramos, his brother, and his cousin and charged them with misdemeanor assault. The crew leaders were sentenced to one year of probation and were required to pay Martinez for the replacement of his vans' windows.

But Germino and her associates were not ready to give up. Unfortunately, workers around Lake Placid refused to talk to the coalition out of fear. So coalition member Romeo Ramirez volunteered to go undercover as a Ramos crew member. Short, even by the standards of his ethnic group, and only nineteen years old at the time, Ramirez knew that he would be killed if his cover was blown. "You have to come in as being very humble, innocent about life, and be with these people, be one of them," he said. "You do it *indigena* style. I pretended to know nothing."

Ramirez was housed in a former tavern just across the highway from El Mercadito. He described it as one of the most miserable places he had ever occupied, which for a veteran migrant tomato picker, is saying something. Cockroaches skittered everywhere. The mattresses reeked. The bathroom was worse than the filthiest outhouse. Even though he maintained a low profile and made no overt inquiries, it wasn't long before other crew members began to confide in the quiet new recruit. Many of them, he was told, were being held captive and were receiving no pay. Three workers were particularly desperate. They feared Ramos wanted them killed and told Ramirez they could no longer endure the suffering. He said that he knew of a way they could escape, but it would carry considerable risk. They were willing to accept that, and Ramirez called his colleagues in Immokalee. A plan was devised.

Benitez drove into Lake Placid just before dark on the appointed evening. He pulled over to the shoulder in front of the former tavern where the men were housed, got out of his car, raised the hood, and pretended to fiddle with the engine. Meanwhile, the balcony of a motel about a hundred yards away afforded Germino and her husband, Greg Asbed, also a coalition member, a clear view of what was transpiring beside the highway. Seeing no other traffic and no suspicious parked vehicles, they signaled. At first, the three workers walked casually toward Benitez's car, but halfway there they panicked and broke into a run, jumping into the backseat and lying on the floorboards. Benitez slammed down the hood, got behind the wheel, and tromped on the accelerator.

Ramiro Ramos, his brother Juan Ramos, and his cousin Jose Luis Ramos were subsequently tried on numerous counts of conspiring to hold people in involuntary servitude, using a firearm during the commission of a violent act, and harboring illegal aliens for the purpose of financial gain. Ramiro and Juan received twelve-year federal sentences; Jose got ten.

Because of the crucial role played by victims and witnesses in the Flores case, congressional legislators inserted a clause in the Victims of Trafficking and Violence Protection Act allowing victims of slavery who cooperate with law enforcement officials to receive T-1 visas, documents that allow them to stay and work in the country for four years and can lead to permanent residency. In the Navarrete case, Lucas Domingo and his fellow slaves agreed to testify against the family that had brutalized them in return for T-1 visas, even though they feared for their lives, especially if the brothers were acquitted.

Early on, Jose Navarrete broke down and pleaded guilty to five of the less serious charges involving harboring aliens. A court date for the five remaining Navarretes was set for September 2008. But a day before the trial was to begin, all five pleaded guilty. "In federal court, if you go to trial and lose, the sentences are extremely severe," defense attorney Joseph Viacava said. "We were happy to negotiate a resolution that caps our clients' liability and puts him in a favorable position come sentencing."

That day arrived in late December, just over a year from the early morning when Domingo had broken through the ventilation shaft in the produce truck. Some former slaves have described the experience of being freed as coming "out of the darkness and into the light," or "from death back to life." Testifying at the sentencing hearing, Domingo told Judge John E. Steele, "Bosses should not beat up people who work for them." As Domingo spoke, Geovanni Navarrete shook his head and curled his lip in a contemptuous smile. It was soon wiped away. Cesar and Geovanni Navarrete received jail sentences of twelve years. Ismael, who pleaded guilty to lesser offenses, was jailed for three years and ten months. The family matriarch, Villhina, was deported to Mexico. Court documents revealed that over the years the Navarretes had deprived their workers of nearly $240,000 in wages.

Some crew members were deprived of something more important than money. Medel, who is in his late forties and goes by the name Don Pacito due to his short stature and regal facial features, was not able to get his life back together. When I spoke to him two years after the verdict, he was homeless, living in a makeshift camp in the woods outside Immokalee. He had been mugged and robbed three times. The drinking problem that had been encouraged by the Navarretes had gotten worse, if anything. He lived in constant fear because friends and relatives of his captors were still around town, many of them running work crews. He rarely walked anywhere alone. It was tough to pick up jobs because many crew leaders refused to hire a known snitch. He was afraid to work in fields in which Navarretes' friends might also be working, so even if he was offered a job, he often refused to take it. When we spoke, he was hopeful that he would soon get enough money together to afford space in one of Immokalee's shabby trailers.

When the verdicts were handed down, there were no howls of outrage from growers. On the contrary, a few weeks after the Navarretes pleaded guilty, the Florida Fruit and Vegetable Association, a trade group, awarded a farmer named Frank Johns its Distinguished Service Award, even though Johns's company operated a farm where slaves freed in an earlier case had been forced to work. Two of the corporations on whose land the Navarrete crew worked, Six L's and Pacific Tomato Growers, did not comment. The only official word from the industry came from Reggie Brown, executive vice president of the Florida Tomato Growers Exchange. "We abhor slavery and do everything we can to prevent it," he said. "We want to make sure that we always foster a work environment free from hazard, intimidation, harassment, and violence." Growers, he said, cooperated fully with law-enforcement officers in the Navarrete case. Charlie Crist, then governor of the state, refused to meet with the coalition to discuss issues related to slavery and labor abuse. Terence McElroy of the Florida Department of Agriculture and Consumer Services inadvertently showed the lack of respect that Tallahassee politicians and bureaucrats have for the people who toil in one of the state's most valuable industries when he downplayed the Navarrete verdict by saying, "Any instance is too many, and any legitimate grower certainly does not engage in that activity, but you're talking about maybe one case a year."

Maybe he should have talked to attorney Molloy or Detective Frost, neither of whom had time to savor victory in what had been one of the most brutal, vicious slavery cases yet prosecuted. Both were too busy trying to put together enough evidence in two other human trafficking cases as McElroy spoke. A few months later, when I visited Frost in his Naples office and asked him if he thought there were other workers being held in conditions similar to the Navarrete crew, he pointed due east, toward the fields of Immokalee, and said, "It's happening out there right now."

Viacava, the Navarretes' defense attorney, offered what is perhaps the most salient last word on the case, saying that it exemplified the hypocrisy of the U.S. legal system. The original version of the Victims of Trafficking and Violence Protection Act included language that would have made it possible to jail those who profit by "knowing or having reason to know" that workers under their ultimate control were enslaved. That would have included the executives of the large tomato companies. According to testimony by the coalition's Benitez at a hearing of the Inter-American Commission on Human Rights, just as it seemed like the bill would pass easily with that language, Senator Orrin Hatch (R-UT) insisted that the clause be removed from the act. "We have a migrant worker being prosecuted to the fullest extent of the law with all the government's resources while these multimillion-dollar corporations stay off in the distance," Viacava said. "If the corporations are going to employ these illegal migrant workers, they should be equally responsible. If you want to truly cure these ills, go after them. But I don't think that's going to happen—my clients don't have the ability to make huge campaign contributions."

As for Lucas Mariano Domingo, after his day in court, he took his new visa and went back into the same fields where he had once been a slave, still hoping to make enough money to send home to his ailing parent.

AN UNFAIR FIGHT

With four thousand members, the Coalition of Immokalee Workers occupies a prosperous-looking, coral-colored, one-story building that provides meeting and office space. It is also home to a radio station that the coalition runs and a small general store stocked with the basics of a Hispanic kitchen priced low to dissuade other shopkeepers in town from gouging customers. I met Geraldo Reyes there one afternoon. He had agreed to guide me on a walking tour of the migrants' side of Immokalee, a part of town few outsiders see.

Reyes is a tall, lanky Mexican in his early thirties. His normal demeanor is sleepy eyed and serious, but it can lighten instantly when he breaks into his infectious gap-toothed smile. He came to the United States about fifteen years ago and rented mattress space in a trailer. One of his roommates told a harrowing tale of being held as a slave and how a group called the Coalition of Immokalee Workers had helped set him free. Intrigued, Reyes started attending weekly discussion group meetings organized by the coalition and eventually began working for the group almost full time, although he still puts in a stint as a watermelon picker in North Florida and Georgia each summer.

As we set out along a sandy path, he told me that it was wrong to view slavery in Florida's fields as a series of isolated cases. Rather, he explained, slavery is an inherent part of an economic system built on the ruthless exploitation of its workers. In this grim continuum, there is not much difference between an actual slave and a man who, say, has put his family's property in Mexico up as collateral for a loan from an unscrupulous crew boss to get across the border to Florida and who must work indefinitely just to pay off that loan. A tiny step beyond that along the continuum is the worker who may not be indebted to his boss but has to pay him inflated rates for lodging, transportation, and food. Mary Bauer, who represents migrant agricultural workers at the Southern Poverty Law Center, said, "There are these really terrible, dramatic slavery examples, and then there are less dramatic, but still incredible oppressive circumstances that, in effect, amount to forced labor that are extremely common and, in fact, close to the norm."

Federal labor laws helped create and continue to support this abusive economy by specifically denying farmhands rights that virtually all other American workers take for granted. As part of Franklin D. Roosevelt's Depression-era New Deal, the National Labor Relations Act granted workers the right to form unions and engage in collective bargaining without fear of being fired. In early drafts, the legislation covered everybody, but the final version exempted domestic help and farmworkers from the basic protections provided by the act. In the wording of the bill, the definition of "employee" did not include "any individual employed as an agricultural laborer." The official reason given was that, in those days, farmers kept at most only one or two hired men and that households had only a few domestic servants, so unions and collective bargaining were not an issue. But in the 1930s, most domestics and farmworkers in this country were African American, and Roosevelt needed the support of southern Democratic legislators to get his New Deal legislation through Congress. Legalized discrimination against farmworkers is not limited to the Labor Relations Act. The people who plant and harvest our food are also exempt from laws mandating benefits, and they receive no guaranteed overtime, even if they put in more than eight hours in a day or forty hours over a week. Children as young as twelve years old are allowed to do farm work. In other industries, child labor laws set the minimum age at sixteen years.

The thirty thousand tomato harvesters who work in Florida are typically paid at least partially on an antiquated "piece basis," meaning they receive a set amount of money for every container of fruit they pick. Technically, the law says that what they are paid must equal at least the minimum wage, which in Florida is $7.25 an hour. Under ideal conditions, a good, hardworking harvester should be able to pick the ten or so bushel-size buckets required to earn that amount in an hour. The problem is that conditions are rarely ideal, and a lot can go wrong. Farm laborers have to be on call every day of the week in case there's work to be done. But if it rains, they can't pick. If their crew gets to the worksite and the vines are covered in dew, they wait unpaid until the vines dry. The trip from town to the farm on the crew boss's bus can take hours, and they receive no pay for travel time. If trucks are not available to transport the harvest to the packinghouse, the workers wait until the trucks are available. For an hour at the end of every workday, they sit around earning no money while the boss tallies the amount each member of his crew picked that day. And the system is plagued by fraud. Bauer of the Southern Poverty Law Center said that her organization has prosecuted numerous cases where field bosses falsely recorded fewer hours on time sheets than a crew member actually worked.

Tomato workers get no sick benefits and no paid vacation. If they are hurt on the job—serious back injuries are common under stoop labor conditions—they have to pay their own medical bills, if they can afford to see a doctor at all. "You have to work like a freak to make enough money so your family can eat," one coalition member told me. "If this was a normally paced, decent job, a lot of the injuries wouldn't happen." This might explain why the life expectancy of a migrant worker in the United States is only forty-nine years. According to U.S. Labor Department figures, migrant workers typically make between $10,000 and $12,000 a year, a figure that is distorted because it includes the higher wages paid to field supervisors. Based on forty-hour work weeks, that means workers' hourly earnings are between five and six dollars, well below minimum wage. The average household income for farmworkers in the United States is between $15,000 and $17,500 a year, well below the federal poverty line of $20,650 and less than half of what is considered a living wage for someone residing in Immokalee. "Most people hope to come here and send money home and perhaps make enough to return there someday," Reyes said. "But when you get here, it's all you can do to keep yourself alive with rent, transportation, and food. Poverty and misery are the perfect recipe for slavery."

Reyes introduced me to a worker named Emilio Galindo. I asked Galindo to describe a typical day in the life of a tomato worker. Originally from Puebla, Mexico, where he once earned a living for his wife and four children making bricks and growing a bit of corn, Galindo was in his early fifties, giving him the status of a senior citizen among pickers, who are mostly in their twenties and thirties. He was a short, stocky man built like a compact bear, and he favored his right leg as he walked. Two tufts of gray hair stuck out from beneath his ball cap. Galindo said he had been harvesting tomatoes in Immokalee for ten or eleven years—he couldn't recall the exact number but said he had managed to get home only once during that period. His days start at four o'clock in the morning to give him enough time to walk through the dark streets to "The Pantry," arriving between 4:10 and 4:30. The Pantry was the name of a store that had closed, but the workers still use the name to refer to the store's parking lot, where they assemble each morning hoping to get hired by one of the crew bosses who stop their crudely painted retired school buses there to load up a team for the day's work.

It is common for Florida tomato farmers to subcontract the actual field work out to crew bosses. Although the company issues paychecks directly to the workers in most cases, the boss is responsible for picking them up, delivering them to the worksite, and then making sure the work is done correctly while they are there. Crews range in size from a dozen to several hundred, but typically consist of between thirty and one hundred pickers. This subcontractor system enables a corporate farmer to avoid direct responsibility for day-to-day abuses that occur in his fields.

Whenever a bus arrives at The Pantry, Galindo said, workers scramble to join the scrum of job hunters clustered near its door. They sometimes jostle to get the best positions, calling out, "Pick me! Pick me!" The leader looks over the group as a grocery shopper might examine a display of tomatoes or a farmer a herd of cattle, selecting the most promising individuals one by one, usually going for the younger, stronger-looking ones first. Once a boss fills his bus, he drives off, and the remaining workers go over to the next bus. Because of his age, Galindo often has to wait for over an hour to get selected, if he is chosen that day at all. Once on a bus, he never knows how far he is going to have to travel to get to the field. "Sometimes fifteen minutes, sometimes two hours," he said. There is often another wait at the field. "You can't go into the fields until the bosses tell you to."

In the fields, the crew boss and his assistants show workers where to pick. If a worker is lucky, or a favorite of the boss, he gets stationed near the trucks that are being loaded. If not, he may be a hundred yards away. "That's the difference between earning $40 and $55 that day," said Galindo. He said that it takes him an average of about five minutes to fill a bucket if there are lots of tomatoes to pick, but sometimes much longer when he has to cover two or three hundred feet of a row to fill a container.

He assumed a crouch in front of me, like a baseball catcher's and gestured toward an imaginary bucket between his legs, making pawing motions with his hands, miming the action of picking. "Your knees hurt the most," he said. "Then your legs and your back." He spread out his fingers. The cuticles were cut, bleeding, and stained black from "tomato tar," a combination of plant resin, dirt, and, he said, "*pesticidas.*" All day long, the boss and his assistants, who are paid on the basis of how much their crews pick, stand over the workers, urging them on, incessantly yelling and swearing, "Hurry! We have to fill two trucks today. Hurry!"

Depending on how work is going, Galindo is sometimes allowed to stop for lunch, sometimes not. The same policy applies to bathroom breaks. Sometimes a boss says it's okay for a worker to trot over to one of the portable outhouses required in the fields, but sometimes—Galindo clutched his bowels and grimaced. "Your stomach hurts and you have to run into some brush bordering the field." At the end of a ten- to twelve-hour day, he said, "You feel all used up." When Galindo gets home, he is usually too exhausted to do anything except fall asleep until four o'clock rolls around again. "We do this because we have to provide for our families," he said. "I thought things would be so much better here. Now, I sometimes think it would be better to have stayed back home, even if all we had to eat was beans."

Reyes and I proceeded through the neighborhood of trailers on bare sandy lots covered in discarded, rotting mattresses, rusting immobilized cars and vans, broken beer bottles, and plastic shopping bags. Leaning wearily against the railing of a wooden stoop in front of one of about a dozen trailers, all painted the same putty color, Juan Dominguez complained to Reyes that he'd had a bad day. It was still early in the season, and the crops were coming in slowly. A boss who needed help planting a field took him on, which should have been a good job. Planters are paid a flat hourly wage, albeit the minimum wage. But when he got to the field, the truck that was supposed to deliver the trays of seedlings from the greenhouse had not shown up. When it finally did, they were able to plant for only a couple of hours, returning to town at three o'clock in the afternoon. Altogether, he had been away from home for nine hours that day. Dominguez's total earnings were just $13.76.

I asked Dominguez if he would mind showing me around the singlewide he and nine other workers rented. He shrugged tiredly and opened an aluminum door that no longer had a screen. The smell walloped me: Not quite body odor, not the stench of cooking or garbage, it was heavy, sweetish, thick, and stale. Unprofessional carpenters had added some extra partitions to the interior and paneled the walls in cheap particleboard that was painted dark brown, adding to the oppressive atmosphere. Dominguez swept his hand in a gesture of invitation into a bedroom. It housed five twin-bed mattresses. Three were flat on the floor with no space between them. Two rested on four-by-eight-feet plywood sheets suspended from the ceiling on chains. The room was covered in T-shirts, jeans, ball caps, running shoes, and a collection of cheap backpacks and luggage. The bathroom was at the end of a short hallway. Barely bigger than an airplane lavatory with a curtainless metal shower stall, it served ten men who came home each day hot, dirty, and anxious to bathe. The sink was stained black. The toilet lacked a seat. The kitchen consisted of a Formica-topped table and four mismatched plastic-upholstered chairs with grayish stuffing protruding from slashes. A saucepan containing something brown and hard rested on one of the burners of an apartment-size stove. A stainless steel sink was set into a counter that no longer had drawers or cupboard doors. A steady dribble of water ran from the faucet, and the door to the badly rusted refrigerator would not close. A single bulb dangled from a cord attached to an open electrical box in the ceiling, and two fans waged a noisy but futile battle against the heat and humidity. In a region where the temperature can soar into the nineties and plunge into the twenties, the trailer had neither air conditioner nor heater. When Reyes and I left the trailer, the sultry air outside seemed fresh and crisp. He shook his head. "You would never live like that at home," he muttered. Yet Dominguez and his housemates paid $2,000 a month for their squalid accommodations, about the same amount as you would pay for a clean little two-bedroom unit in Naples.

Pascuala Sanchez and her three children (four counting the unborn fetus she carried) were among ten farmworkers who lived in a trailer much like the one occupied by Dominguez and paid a similar rent. To thwart would-be burglars, the trailer's occupants had nailed wire mesh over the windows, a common crime-prevention practice in Immokalee. At about 2:30 one Sunday morning in March 2007, fire broke out in the crowded dwelling. In less than one minute, the entire structure was ablaze. Firefighters who responded were unable to get inside because of the mesh; occupants could not escape through the windows for the same reason. Sanchez and two of her children, twelve-year-old daughter Luciana and six-year-old son Rodrigo, along with two other occupants, Emiliano Figaroa and Adelmo Ramos, died. The five survivors were all hospitalized with serious injuries. For a time, neighbors thought that Sanchez's eldest child, Wilder, had also perished. It took more than two weeks for authorities to determine that he was at a hospital in Tampa being treated for third-degree burns to his back, chest, and arms.

In most other communities, a disaster of that magnitude would have sparked demands for immediate improvements in zoning laws. But it changed absolutely nothing in Immokalee, where one-quarter of the residences are substandard, according to county housing officials. After touring Immokalee in 2008, Senator Bernie Sanders (I-VT) described the housing conditions there as "deplorable" and said that the shacks and trailers would never have passed a safety inspection in Burlington, the small Vermont city where he had once been mayor. To give them credit, community leaders who want to improve housing in Immokalee find themselves in a catch-22. Field workers need places to live. The sort of aggressive enforcement of building codes needed to bring the housing in Immokalee up to standard would dump hundreds of workers on the streets. With fewer spaces available, slumlords could then charge even more inflated rents for those that remained, and a highly profitable racket would become even more lucrative. The easy money to be made renting shacks to migrant workers at Manhattan prices exacerbates a problem government housing departments and charitable groups face in acquiring land on which to build decent, affordable housing in the area. Owners demand outrageously high prices to sell the land, based on the returns they are earning—a dozen trailers on a single city lot can generate annual revenues of more than $200,000, most of it profit, making that lot worth well over $1 million as an investment. County officials did attempt to shutter a thirty-four-unit trailer park owned by Jerry and Kimberlee Blocker, members of an extended family that controls many rental properties in Immokalee, because it deemed the structures uninhabitable. The couple promptly sued. They maintained that the dwellings were completely habitable and that in any case, the park dated back to the 1940s, long before the county's zoning laws were enacted. Their lawyer, Margaret Cooper, said that her clients wanted the court to declare that the trailer park was grandfathered. "They have vested rights under the prior laws," she said.

One of the reasons that rents are so high in Immokalee is that many workers lack vehicles and must live within walking distance of the downtown pick-up areas where crew leaders' buses stop each morning and evening. The busiest of these is a football-field-size parking lot in front of La Fiesta, a sprawling building housing a supermarket, taqueria, deli, and check-cashing outfit. I arrived there a little before five o'clock in the morning, which would have made me a slacker among tomato workers. The place was already bustling. A dozen amateurishly repainted school buses with hand lettering on their sides saying "Montano Harvesting" or "A. and J. Field Services" sat in the yellowish glare of the streetlights. Several women stood beside the opened backs of SUVs selling tacos and tamales to workers, who appeared out of the shadows along a web of sandy paths leading from the trailers. The scene was eerily quiet, except when the local population of roosters erupted in vigorous crowing contests. I stopped at a group of about ten guys sitting on top of a picnic table. One of them told me that they waited there every day for a crew leader who had hired them for the season. As regular members of a crew, these workers represented the highest social class in that predawn gathering. But hundreds of other men with no certain prospects had lined up. Their only hope for work was if a crew was shorthanded or a farmer was in desperate need of a group of temporary pickers. This group was the lowest of the low—the bottom of the bottom of the American labor force. By seven o'clock, the sky began to lighten. The number of buses coming and going slowed, and then stopped. Several dozen men still milled around the lot, hoping that a crew leader might have been delayed or would be summoned at the last moment to harvest a field. But it soon became apparent that there would be no more buses that morning. The men straggled off toward the trailers, thermoses and plastic grocery bags full of lunch slung over their slumped shoulders, shuffling their feet even more wearily than the workers who got off the buses that evening after a full day in the fields.

Later that same morning, I saw a few of the men who had gone away without work when I volunteered to put in a shift at the Guadalupe Center soup kitchen. The Guadalupe Center is a charitable organization whose mission is to "serve the migrant and rural poor of Immokalee." It operates a clothing room, where donated garments, toys, and small appliances are sold at a rate of one dollar per full shopping bag—a fee put in place when the organization found that paying a dollar allowed customers to maintain their pride and increased use of the room. It also runs a shower program, providing fresh towels, clean clothes, and toiletries so that workers without access to plumbing can maintain their hygiene as well as their dignity. It gives out five hundred pairs of new shoes to area children each August just before the beginning of the school year. And it runs a daycare and preschool program. The center was started in the early 1980s when some volunteers and clergy members remodeled a building owned by the Catholic church of Immokalee for use as a soup kitchen, which, by the time I tied on an apron, had served more than one million free hot lunches.

Tricia Yeggy, a high-energy young blond woman who was then the director of the kitchen, explained that the place runs on two straightforward rules: People can eat as much as they want, and no one is turned away hungry. That meant serving around three hundred meals a day in seatings of forty-five diners each that began at eleven o'clock and ran until there was no one else waiting. Yeggy pointed me toward a collection of jugs filled with orange juice and told me to hurry along and put two on each table. My fellow volunteers that morning were all retirees from a Naples church group and veterans of the soup kitchen. Everyone knew the drill. They bustled about, cutting bread, setting tables, putting out bottles of hot sauce. When the first sitting of "guests," as the kitchen's clients are called, came in, we loaded up trays with bowls of turkey and rice soup, thick with summer squash, corn, and a vigorous sprinkle of cumin. It was both hearty and tasty. If a guest finished one bowl and wanted another, he would raise his empty bowl overhead, and one of the volunteers would rush over with a full replacement. The dining room was decorated with white lace curtains and plastic tablecloths bearing colorful floral designs or images of Campbell's soup cans à la Andy Warhol. Walls were painted in bright shades of pink, purple, and orange. The volunteers were cheerful, the guests uniformly grateful, smiling shyly, saying, "*Gracias*." At shift's end, I almost forgot the underlying irony: Workers who pick the food we eat cannot afford to feed themselves.

Large picture windows of the coalition building overlook La Fiesta's parking lot from across the street. The highly visible location was chosen to provide easy access to workers and, I suspect, to serve as a constant reminder to crew bosses that someone may be watching. One morning, I encountered Lucas Benitez, sitting alone at his desk in the building. Benitez, who had driven the get-away car to free the workers held by the Ramos family, is a paunchy man in his early forties who wears a severe flat-top hairdo and a trim goatee. The main spokesman of the coalition, he was one of six children in a family from the southwestern Mexican state of Guerrero. He came to the United States at seventeen to help support his parents and siblings. One day early on in his work in the fields, he was driving stakes to support tomato plants. Fit and young, he soon got well ahead of other members of his crew and stopped briefly while they caught up. A boss started yelling at him and when that had no effect got out of his pickup truck, saying he was going to beat Benitez. The other crew members turned their backs, or looked down. Benitez realized that he was alone in the middle of thousands of acres of fields. Nonetheless, he brandished a tomato stake and faced the boss down—that day. He soon began meeting with a small group of workers during the evenings in a room offered by Immokalee's Catholic church. They discussed their poverty and the brutal conditions they worked under and decided that if they worked together, no longer looking down or turning away from abuse, they could improve their conditions. That group became the Coalition of Immokalee Workers.

I asked about the organization's early activism. Benitez didn't answer. Instead he went over to a stack of filing cabinets and opened and closed one after another until he found what looked to me like a crumpled rag. He sat back down and gingerly unfolded it on the desk in front of me. It was a faded blue-and-white-striped shirt made of a coarse, canvaslike material. It bore the telltale black stains of "tomato tar" but was also covered in brownish splotches —dried blood. "This is Edgar's shirt," Benitez said.

Edgar was a sixteen-year-old Guatemalan boy who staggered into the coalition's office one afternoon seeking medical help. He was covered in blood and told a story of becoming thirsty while out in the fields. When he asked for water, his boss told him to shut up and keep picking. Overcome, the boy stopped long enough for a drink. His crew leader bludgeoned him until Edgar managed to get away and run for his life. A few members of the coalition confronted the boss later that afternoon, but he snorted in derision, knowing damn well that there wasn't anything they could do.

Later that night, nearly two hundred workers attended a meeting in the coalition's offices. They decided to march to the boss's house. As they proceeded, others came out of the work camps and trailer parks, swelling the throng to more than six hundred by the time they arrived at their destination. Twenty-eight police vehicles and a brigade of cops in full riot gear were waiting. The leaders of the march brandished the bloody shirt, chanting, "This shirt is Edgar's. It might be mine next. When you beat one of us, you beat us all." The next morning, when the crew boss who had beaten Edgar pulled up to the parking lot in his bus, not a soul would get aboard. Other crew leaders took note. That happened in 1996. "It was the last report we got of a worker being beaten by his boss in the field," Benitez said.

Equally important, it marked a turning point. Workers saw that by organizing, they could effect change. The shirt, now framed and hanging in the Florida Modern-Day Slavery Museum, is kept by the coalition as an unofficial flag, a reminder that by banding together, they can make progress. In some cases, "progress" means helping individuals. Romeo Ramirez was the same age as Edgar when he first reached out to the coalition. His boss had shorted his paycheck. Noticing that his employer was preparing to leave town for the annual journey north, Ramirez asked some acquaintances what he should do. "Maybe the coalition can help you," they said. The coalition arranged for twenty-four workers to get into two vans and pay a visit to the boss. Just the sight of them was all the encouragement he needed to pull out his checkbook and settle up with Ramirez, who now works for the coalition. "They helped me," he said. "It seemed right for me to stay and help the next guy." It was Ramirez who risked his life two years later by going undercover to infiltrate the Ramos slavery gang, freeing several hundred fellow migrant workers.

In the early days, progress was stop-and-go. Six coalition members staged a month-long hunger strike that began in December 1997, after ten major growers ignored a request signed by nearly two thousand area farmworkers to meet with the coalition to discuss wages, which had fallen from fifty cents a bucket to forty cents over the previous two decades. As the New Year started, the hunger strikers began to weaken. One required hospitalization. Prominent clergymen came to visit them on the cots where they lay, taking only water and juice. Governor Lawton Chiles urged the growers to "begin a meaningful dialogue with representatives of these workers." Even that did not move the farm owners. When one of them was asked why he refused to listen to the worker's requests, he replied, "I'll put it to you this way. The tractor doesn't tell the farmer how to farm." In an early flash of the savvy public relations stunts that have become key to the coalition's success, the next time the workers held a demonstration, they wore white headbands with the words *Yo no soy tractor* ("I am not a tractor") printed on them in red.

Ultimately, one grower sat down with the coalition and agreed to give his workers a ten-cent-a-bucket raise. Others followed. For all their efforts, the coalition had managed to bring pay rates back only to what they had been in the 1970s, not factoring in inflation. But at least the decline had stopped. When pressed for more money, the growers cried poverty. They faced competition from Mexican farmers, who paid even lower wages. The huge fast food companies that were some of their biggest customers would simply go south of the border for cheaper tomatoes. At one of the coalition's regular weekly meetings, a member—nobody recalls exactly who—came up with what amounted to an end run around the farmers' arguments. If the farmers say they can't pay us more, why not take our case directly to their fast food customers? It was a masterful public relations gambit. In the eyes of consumers, the tomatoes on their hamburgers or in their tacos are a faceless commodity. Nobody knows the name of the corporate farm that grew and packed them, so the growers could ignore public opinion. But fast food companies that spend millions of dollars cultivating a wholesome, family-loving reputation with smiling clowns, cute red-haired girls with pigtails, and grandfatherly old gents in white suits could not afford to have their brands linked to images of abused farmworkers. The coalition devised what they called the Campaign for Fair Food. In another coup, the workers' request would center on something every consumer could relate to: Guarantee us a few basic rights and give us one penny more per pound for the tomatoes we pick. A penny per pound would be a pittance to a fast food behemoth like McDonald's, which has annual revenues of over $22 billion. But when you are picking a ton of tomatoes a day, as a worker typically does, that's a raise from fifty dollars a day to seventy, the difference between below-poverty existence and a livable, if paltry, wage. The coalition singled out its first target: Taco Bell, a company that had built its brand on television advertisements starring a sombrero-wearing, Mexican-accented Chihuahua named Gidget. The advertisements had already offended many Hispanics and other politically sensitive viewers who saw them as a crass (albeit hugely successful) attempt to perpetrate and commercialize a racist stereotype of Mexican culture. The members soon found out that taking on the world's largest fast food empire (Taco Bell is owned by Yum! Brands, a 35,000-restaurant chain that also controls Pizza Hut, KFC, A&W, and Long John Silver's) was going to require a much more sophisticated effort than staging a protest march or rounding up a couple of dozen members to pay a collection call on a recalcitrant crew boss.

How was a grassroots organization with a limited budget, based in a poor backwater in South Florida, whose members could barely afford dietary basics, going to convince corporate executives in suburban Los Angeles to forgo profits by volunteering to pay a penny per pound more for the tomatoes that went into their salsa and salads? The answer was to hit them where it hurt the most—in the bottom line. In 2001, tearing a page from the successful grape boycotts mounted by Cesar Chavez and the United Farm Workers in the second half of the 1960s, the coalition launched a boycott of Taco Bell.

From the outset, the coalition knew it would need all the allies it could get. One obvious place to look was the nation's college and university campuses. Taco Bell's core consumer target group was eighteen- to twenty-four-year-olds, whom the company cynically called the "New Hedonism Generation." But the coalition saw entirely different traits in young people. They believed that the college students had shown that they felt deeply about social justice and would take action to bring it about, whether it meant refusing to purchase logo-emblazoned clothing produced in Asian sweatshops, supporting unionization efforts of blue-collar campus workers, or battling administrators intent on cutting back academic budgets while padding their own salaries and enlarging their staffs. Activist groups were already in place on most campuses and, better yet, were adept at communicating through the Internet and other new media. The coalition's Web site became a crucial new tool in its battle for fair food.

In 2000 the coalition decided to sponsor a 230-mile protest march that culminated at the Florida governor's mansion in Tallahassee. A couple of dozen college students accompanied the marchers to collect signatures on a petition that would be delivered to then Governor Jeb Bush. After the demonstration, the students formed the Student/Farmworker Alliance to put an end to what they called "sweatshops in the fields." In mid-2001, when the Taco Bell boycott was announced, the alliance had branches at three Florida universities, but through the Internet, had established relationships with student groups at every major post-secondary institution in the state. Within a few years, that number grew to more than three hundred colleges and universities in all parts of the country, including the University of California Los Angeles, the University of Chicago, and the University of Notre Dame. The alliance also established ties with activist groups in fifty high schools. In 2004 hundreds of those students went on hunger strikes to "Boot the Bell" off their campuses, and in twenty-two cases, the schools did just that. Taco Bell managers learned the hard way that ignoring the rag-tag workers coalition would carry a price tag, but they still stubbornly refused to accede to the coalition's demands.

The struggle for workers' rights can make for strange bedfellows, and while the students' campaign was gaining momentum, the coalition reached out to religious leaders through a group called Interfaith Action of Southwest Florida. Eventually it gained the support of the Presbyterian Church (U.S.A.) and the National Council of Churches, which speaks for forty-five million parishioners in more than 100,000 local congregations in the United States. With lodging and logistical assistance from the church groups, the coalition was able to mount a cross-country Taco Bell Truth Tour in 2004 that culminated in a massive demonstration and a ten-day hunger strike outside Taco Bell headquarters in Irvine, California.

While the students and farmworkers continued their highly visible campaign against Taco Bell in the streets and media, religious folks exerted pressure more quietly but no less effectively in the rarified atmosphere of the boardroom. At a 2003 shareholders' meeting of Yum! Brands, Taco Bell's parent company, a representative of Oxfam America presented a resolution calling for Yum! to undertake a transparent review of its policies related to social and economic sustainability throughout its supply chains. The resolution was clearly aimed at the conglomerate's failure to respond to the Immokalee workers' requests. Such resolutions are frequently put forward at annual meetings of major corporations. Typically, they get less than 10 percent of the votes and managers shrug them off as the feeble cries of extremists and kooks. The Oxfam resolution won the support of more than 35 percent of the shareholders. There was no way the executives could ignore that result. By early 2005, Yum! had agreed to the coalition's requests, not only for its Taco Bell restaurants but for every restaurant in the conglomerate.

Two years later, McDonald's signed a similar agreement. But the coalition was making little headway against Burger King, a situation that was all the more painful because the chain was founded in Miami in the 1950s and was still headquartered there, less than one hundred miles from the fields of Immokalee. If any group of fast food executives should have been aware of the plight of the people who picked their tomatoes, it was the officers who ran the world's second largest hamburger chain. Instead, they mounted aggressive resistance to the overtures from the workers' representatives. It was an approach that backfired badly.

In 2007 and 2008, a writer who went by the Internet pseudonym surfxaholic36 frequently attached ungrammatical but scathing comments to articles, blog postings, and YouTube videos mentioning the Coalition of Immokalee Workers. A typical one read, "The CIW is an attack organization lining the leaders pockets . . . They make up issues and collect money from dupes that believe their story To bad the people protesting don't have a clue regarding the facts. A bunch of fools!" In a reply to a story in the *Naples News* that covered a visit to Immokalee by Senator Bernie Sanders (I-VT), surfxaholic36 wrote: "The CIW is an attack organization and will drive business out of Immokalee while they line their own pockets. They make money through donations by attacking large companies and have attacked Yum, McDonald's and now Burger King to get money for there own organization. They are the lowest form of life exploiting the poor workers to line there own pockets. I will buy all the Whoppers I can, good going Burger King for uncovering these blood suckers."

Amy Bennett Williams, a reporter at the *Fort Myers News-Press*, decided to do a bit of Internet sleuthing to see if she could uncover the true identity of surfxaholic36. It didn't take her long to trace the alias back to Shannon Grover, who lived in the Miami area. Williams put in a call to Grover, and a youthful female voice answered. Williams identified herself, making sure that the girl on the other end of the line knew that she was speaking to a reporter and understood the ramifications. Grover confirmed that she was indeed surfxaholic36, mainly when she visited social networking sites. Williams then asked her about the vitriolic comments, and Grover said, "I don't really know much about the coalition and Burger King stuff. That was my dad.

My dad used to go online with that name and write about them." Williams asked the middle-school-aged girl if she had ever written about the coalition online. "No," she replied adamantly. "That was my dad. That was him." Shannon Grover's dad was Steven F. Grover. At the time, he was vice president of food safety, quality assurance, and regulatory affairs for Burger King. "This is a huge black eye for Burger King. It's the type of situation that lands companies in public relations textbooks on how not to engage the press, the public, and your critics," said John Stauber of the Center for Media and Democracy.

It was not the only textbook-quality public relations flub made by Burger King. In November 2007, shortly before the coalition was scheduled to march to the company's offices to present a petition, the fast food chain teamed up with the Florida Tomato Growers Exchange, an agricultural cooperative representing more than 90 percent of the state's producers, to sponsor a junket for a group of journalists to Immokalee. The idea was to let the reporters see firsthand that all was well in the tomato fields. Burger King took the occasion to present the Redlands Christian Migrant Association with a check for $25,000 to support the network of daycare centers and schools the association runs. "We found an organization we could trust and we could help," Grover said.

"We finally said, 'Enough,'" Reggie Brown, the executive vice president of the Tomato Growers Exchange, told the reporters. "We're not going to be accused of things we don't do. This is certainly not a labor force held in servitude." Brown's assertions were supported by André Raghu, the managing director of Intertek, a firm that conducts audits to ascertain whether suppliers to food-service companies follow proper quality, safety, and regulatory protocol. Raghu said that his investigators had found "no slave labor" in the tomato fields, which was an odd assertion given that a quick Google search would have popped up at least five Florida cases that had been successfully prosecuted by that date. Nonetheless, the journalists dutifully quoted the Tomato Growers Exchange and Burger King. In a seeming coup for the fast food chain and its suppliers, the *Miami Herald* ran a lengthy article detailing what the reporters had heard and seen. It ran on November 20, 2007.

The timing could not have been worse. That article appeared on the very day that the workers for the Navarrete family escaped from the locked truck in Immokalee and reported to police that more than a dozen migrants had worked as slaves in one of the most heinous cases of human trafficking ever prosecuted in Florida. Finally, in mid-2008, John Chidsey, chief executive officer of Burger King, agreed to join the Campaign for Fair Food. "We are pleased to now be working together with the CIW to further the common goal of improving Florida tomato farmworkers' wages, working conditions and lives," he said at a signing ceremony:

> The CIW has been at the forefront of efforts to improve farm labor conditions, exposing abuses and driving socially responsible purchasing and work practices in the Florida tomato fields. We apologize for any negative statements about the CIW or its motives previously attributed to BKC or its employees and now realize that those statements were wrong. Today we turn a new page in our relationship and begin a new chapter of real progress for Florida farm workers.
>
> For more than 50 years, BKC has been a proud purchaser and supporter of the Florida tomato industry. However, if the Florida tomato industry is to be sustainable long-term, it must become more socially responsible. We, along with other industry leaders, recognize that the Florida tomato harvesters are in need of better wages, working conditions and respect for the hard work they do. And we look forward to working with the CIW in the pursuit of these necessary improvements. We also encourage other purchasers and growers of Florida tomatoes to engage in dialogue with the CIW in support of driving industry-wide socially responsible change.

Steven Grover was no longer with the company.

One March evening, I received an invitation to drop over to the headquarters of the Coalition of Immokalee Workers to take in an amateur boxing match. The turnout was huge. I had to wait in line at the door, and once in, I was relegated to the back on one of several upended plastic tomato buckets that had been hauled out as overflow seating. It didn't look like it was going to be a fair fight. In one corner of a cordoned-off area in front of a meeting hall was a lean, long-armed buff fellow in his late teens or early twenties. He stood well over six feet tall. In the other corner was Emilio Galindo, who had told me about life as a picker earlier. At five-feet-two, he barely came up to his opponent's shoulders.

Galindo's dark blue T-shirt was inscribed in white with three five-pointed stars and the letters *C*, *I*, and *W*. His opponent wore a sign saying "Taco Bell." A referee entered the ring and held up a piece of poster board with "2005" hand-lettered on it. A bell sounded, and the boxers warily approached each other. It was a hell of a bout. They tore into each other in a flurry of puffy red gloves, with each blow being either cheered (if Galindo landed it) or booed (if he was on the receiving end). Things were looking grim for the diminutive Galindo, but he delivered a quick upper cut. His opponent fell to the mat. The fighters returned to their respective corners, where women fanned the victorious Galindo with towels and offered him sips from a water bottle. The bell rang again, and another round ensued. This time the referee held up a poster reading "2007," and Galindo's tall opponent wore a McDonald's sign. After a few minutes of back-and-forth flailing, he was KO'd again. Galindo was just getting warmed up. Burger King came at him in the 2008 round and landed one or two low blows that almost caused Galindo to topple backward. But he managed to stay upright and patiently waited for an opening, which he eventually got. Down went The King.

It was a stunt, of course: *theatro* meant to attract an audience and deliver a simple lesson at one of the coalition's weekly meetings. Like many civil rights groups with Hispanic roots, the coalition embraces the principles of "popular education." Developed in the late 1960s by an exiled Brazilian scholar named Paulo Freire and described in his book *Pedagogy of the Oppressed*, popular education's goal is to empower groups of people who have been marginalized socially and politically. The theory is that once people know where they fit on the social and economic continuum, they can begin taking steps to improve their lot.

Popular education avoids the rigid classroom/teacher hierarchy. A participatory process intended to be an exchange of ideas, it relies heavily on nontraditional means of teaching such as music, visual arts, theater—even faux boxing matches. An illiterate indigenous migrant who speaks no English and only rudimentary Spanish and has just arrived from a hamlet in the mountains of Chiapas may not have heard of Yum! Brands or have any real concept about what a multibillion dollar corporation is, but he can understand the symbolism of a well-liked, plump middle-aged guy in a coalition T-shirt getting whaled by a larger enemy and emerging victorious.

After the boxing match, Benitez came to the front of the room and sat on a desk, spreading his knees comfortably. He glanced out over the crowd of over two hundred—senior citizens, young women, mothers, kids, babies, and many working-age men, their hair oiled back, wearing clean T-shirts and blue jeans and either baseball caps or straw cowboy hats. The room was decorated with souvenirs from the coalition's campaigns. Brightly painted native art murals, banners, and protest signs adorned the walls, along with color photographs of marches and protests. *Yo no soy tractor* headbands hung from window frames like pennants.

In an organization that has no official head (about ten members play key roles in its efforts), Benitez is a natural leader. He is a powerfully built man but speaks quietly. His oratory is soft, conversational, and spiked with low-key humor and good-natured ribbing. He rarely makes a direct statement, instead throwing out questions to his audience, or engaging in one-on-one banter.

"Who was Emilio?" he said, giving his own well-padded belly a few pats that drew a laugh from the crowd. Hands shot up. He pointed to one. "The coalition," came the shouted reply. Benitez swung his head around the room nodding. "And who is the coalition?" he asked. No hands shot up. "Is it Pedro?" he asked, pointing to a worker in the crowd.

The room rumbled, "No."

Indicating another audience member, he said, "Is it Lorenzo?"

"No."

"Is it Lucas?" He pointed to himself.

"No."

Very quietly he asked, "Is it us?"

The room erupted in a cheer.

"When did we start the campaign?" Benitez asked. No one answered. "I think it was in 2001," he said. "What year was it that Taco Bell agreed to become part of it?"

Someone shouted, "2005."

"How many years did it take?"

"Four."

"And what year for McDonald's?"

"2007."

"And how many years was that?"

"Two."

"And what year for Burger King, Subway, and Whole Foods?"

"2008."

By that time, everyone had caught on to his message: The tiny coalition was taking on giant organizations and winning at an increasing pace.

By 2010, the fast food outlets had been joined by Bon Appétit Management Co., Compass Group, Aramark, and Sodexo, all major food-service companies that operate in hospitals, museums, and university campuses. Although major grocery chains (with the notable exception of Whole Foods) still held out, the Campaign for Fair Food was a phenomenal success backed by seemingly unstoppable momentum.

Unfortunately, one huge problem remained. The powerful Florida Tomato Growers Exchange adamantly refused to participate—even though the deals with the fast food and food-service companies would not cost its members a cent. The exchange went so far as to threaten any of its members who did participate with a $100,000 fine. Without the participation of the exchange and big farmers, who maintain employment records for their pickers, there was no practical way to get the extra penny per pound to the workers. Instead of going to people who desperately needed it, the money was building up in escrow accounts, where it benefited no one.

A PENNY PER POUND

"The Coalition of Immokalee Workers has to have a graphic enemy, and I'm it," said Reggie Brown, holding up his palms and flaring his fingers in a what-the-hell gesture. "I cannot get off the hook no matter what—" He was cut off when one of Florida's vicious summer storms unleashed a timely bolt of lightning followed by a sharp crack of thunder.

Technically, Brown has three jobs, though he performs all of them from the same desk in a well-landscaped office park of winding, shaded lanes and low-rise brick buildings just outside Orlando. He is executive vice president of the Florida Tomato Growers Exchange, executive vice president of the Florida Tomato Exchange, and manager of the Florida Tomato Committee. Although the organizations are different legal entities, they share office space and some staff and have overlapping memberships. Observers of the tomato industry could be forgiven for viewing them as a single umbrella organization. Combined, the groups have the power to lobby politicians, advocate on behalf of tomato growers and handlers, advertise and promote Florida tomatoes, fund academic research, impose surcharges on tomato sales, and determine the size and shape of every fresh slicing tomato shipped out of the state during the winter.

Brown is a slim, compact former Marine who still carries himself with a military bearing. He speaks with the soft southern drawl heard in rural North Florida, where his family still runs a farm. Brown is a passionate home gardener, and he gives the impression that he won't be unhappy in a few years when he can leave the politics of tomatoes behind to retire to his plot of land outside Gainesville and tend the small grove of fruit trees he has already planted there in anticipation. When Brown graduated from the University of Florida in 1969, the Vietnam War was in full swing. He was offered a place in the university's vegetable crop PhD program but turned it down and joined the Marines. "Being a typical southerner," he said, "I bit the bullet and did what I needed to do for the country." He left the service in 1973 and started as an extension worker for the Florida agriculture department. He has spent his entire career advising and representing the farming industry through government posts and as an employee of trade associations. As the personification of the Florida tomato industry, he has had occasion to draw on the toughness and discipline he learned in the military forty years ago.

My first glimpse of the power of the Florida Tomato Committee came in 2005, when I encountered a grower named Joe Procacci who was making national headlines by claiming that his company, Procacci Brothers Sales Corporation, had finally cracked the tomato code. Procacci farmed thousands of acres in Florida, and by crossing thick-skinned, disease-resistant Florida field tomatoes with a French heirloom variety called the Marmande, he had managed to breed a good-tasting tomato that was tough enough to be grown in the South in the winter, shipped north, and sold in supermarkets—or so he claimed. Though it might have been good tasting, it was not good looking. Procacci was the first to admit that his new tomatoes, like their heirloom parent, were often asymmetrical, lumpy, and deeply creased. They were so ugly that produce managers often rejected orders, prompting Procacci to make a virtue out of necessity by giving them the unforgettable trade name UglyRipe. They were an immediate hit in the marketplace.

Perhaps too much of a hit. For a few years in the late 1990s and early 2000s, the Tomato Committee allowed Procacci to sell his homely fruits as an experimental crop. But in 2004 when UglyRipes started to become serious rivals to the pretty, smooth-skinned, and utterly tasteless fruits that other farmers grew, the committee ordered him to stop selling them outside the state, even though he had seven hundred acres of ripening UglyRipes in the ground. Procacci had no choice but to feed some of his premium tomatoes to cattle and compost the rest. He lost $3 million. "The cows were eating better tomatoes that winter than the consumers," he said.

Thanks to an arcane document called Federal Marketing Order 966, the Florida Tomato Committee has the ultimate say over the qualities a slicing tomato must have if it is to be exported from the southern part of the state. The Agricultural Marketing Agreement Act of 1937, which paved the way for the creation of Order 966, was passed to allow certain types of farmers to band together and control commodities without being subject to antitrust prosecution. At the time, the act made good sense. In an era when fruits and vegetables were grown by hundreds of small farmers who sold their crops to packinghouses, it assured that growers met consistent standards and that their crops were sold in an orderly manner. Today, when just a dozen large companies are responsible for the vast majority of Florida's production and pack the tomatoes they have grown on their own farms, the financial logic behind such enforced standardization no longer applies. But the power of the marketing orders has in no way diminished. The Florida Tomato Committee decrees the exact size, color, texture, and shape of exported slicing tomatoes. It prevents the shipping of tomatoes that are lopsided, kidney shaped, elongated, angular, ridged, rough, or otherwise "deformed." It delineates down to the millimeter the permissible depth and length of the "growth" cracks surrounding the scar where the fruit has been attached to the stem. It's worth noting that nowhere do the regulations mention taste—it's simply not a consideration. "Taste is subjective," said Steve Jonas, a compliance officer at the committee. UglyRipes failed to meet many of these cosmetic standards. It did not matter that consumers were happy with them and obligingly paid nearly four times what they paid for "Florida rounds," as the gassed mature green tomatoes are called.

By taking on Procacci, the committee had picked a formidable and cagey foe. In his late seventies at the time, he looked like anybody's happily retired Italian grandfather, and in fact a caricature of his smiling face appears on displays of a brand of his tomatoes called Papa Joe's. For all his folksy demeanor, though, Procacci controlled one of the largest produce companies in the United States. Through various corporations, he had interests in Gargiulo, Inc., a major producer of Florida rounds, and also Ag-Mart, the large company responsible for premium niche-market products like Santa Sweets grape tomatoes and UglyRipes. Procacci had been in the business long enough to recognize a once-in-a-life-time marketing opportunity when he saw one, and he was eager to talk. When I met him in the parking lot of a Naples country club, which he and his brother had built on what was once a tomato field, he immediately plopped an inch-thick stack of photocopied press clippings on my lap. They all hewed closely to the same narrative line: A noble farmer grows a great-tasting crop and the Big Bad Tomato Committee won't let him sell it. In reality, it was more like a family squabble among the Goliaths of Tomatoland.

"It's very simple," Procacci explained, as we drove out to see a plot of UglyRipes he had planted to serve the in-state market, which the committee does not regulate. "The committee members are my competitors, and they are jealous. There's a lot of jealousy in this business. They can't have it, so they don't want us to have it, either."

Whatever its motive, the Tomato Committee refused to budge. When negotiations reached a deadlock, Procacci took his case to the USDA, arguing that if the committee allowed producers to export cherry, grape, and plum tomatoes—none of which met its standards for shape—why prevent UglyRipes? The federal bureaucrats turned him down flatly. He brought in lawyers and hired the Washington, DC, lobbying firm that employed John R. Block, who had been President Reagan's agriculture secretary, to take the UglyRipe message to the highest levels in the country.

Procacci did not have to tell me when we arrived at his UglyRipe field. Against a windbreak of cypress trees, I saw several rows of staked vines being ministered to by a harvesting crew. Instead of grabbing the tomatoes off the vines with their bare hands as fast as they could, throwing them into buckets, and unceremoniously upending those into an open truck, these pickers wore gloves to prevent scratching the fruits with their fingernails. They eased the ugly tomatoes two deep into plastic flats. The bottoms of those fruits were blushing a light shade of pink. In a field of Florida rounds, any color other than green is considered seditious. Procacci told me that the first-class treatment of UglyRipes continued at the packing plant, where they were hand sorted (only four out of ten would be deemed worthy to wear an UglyRipe trademark sticker) and individually slipped into foam-mesh "socks" for their journey to grocery stores.

"We can pack fifty or sixty truckloads of mature green tomatoes a day with the same amount of help as we need to pack two truckloads of UglyRipes," said Procacci. "We have to charge a premium price for UglyRipes, but people are willing to pay it. My competitors have all this money invested to process the mature greens, and they want to protect that investment. But it is a diminishing market. More and more, people want flavor. Consumers are not going to eat fruits and vegetables if they don't taste good, and they are going to eat more of them if they do taste good."

The fight between Procacci and the Tomato Committee played out in the U.S. Senate. Senators Arlen Specter and Rick Santorum, Republicans representing Pennsylvania, where Procacci Brothers Sales is based, introduced legislation that would specifically exempt

UglyRipes from federal grading standards, provided that the agriculture department enrolled them in a special program that was designed to track genetically modified foods—which UglyRipes are not. Despite a twenty-four page appeal written by Reggie Brown on behalf of the Tomato Committee and a personal letter from then Florida Governor Jeb Bush, UglyRipes got their exemption in early 2007, days after Governor Bush left office.

Procacci had scored two victories in one. I have eaten my share of $4-a-pound UglyRipes. None have packed anything close to the flavor power of a locally grown summer tomato. Some have been quite pleasant tasting, some so-so, and others not good at all. But no matter. The melee in the press did more to polish UglyRipe's image than the most elaborate advertising campaign ever could have. With his exemption in hand, Procacci was able to cash in on years of pent-up consumer demand. The Tomato Committee had done him an enormous favor.

Reggie Brown represents an industry that faces far greater problems than whether to allow one of its key players to sell homely tomatoes. Almost all American farmers have seen their share of the retail price of their product decline steadily as middlemen gobble up greater margins, but Florida tomato growers have been falling further behind than most. In the last three decades, the price we pay for fresh tomatoes in supermarkets has increased fourfold. During the same period, what farmers receive has only doubled. At the best of times, the business is a high stakes gamble. Growers spend millions of dollars to put in a crop and then have to hope that their plants are not hit by a hurricane or a freeze. Even after a bumper harvest, there is still no guarantee that a grower will be able to sell his crop profitably in a market that is often saturated with tomatoes. Unlike corn, soybeans, and wheat, which can be stored until prices improve, tomatoes are perishable and have to be sold soon after they are picked. "But sometimes you can make money," Brown said.

The summer of 2010, when I met with Brown, was not one of those times. That winter, growers in Florida had been hit by a freeze that destroyed 80 percent of the state's crop. "For every hundred acres of tomatoes that you lost in the freeze, you could kiss $1 million goodbye—gone!" said Brown. Because they had no other choice, farmers replanted the fields, but the weather refused to cooperate. Florida had sixty days of below-normal temperatures in early 2010, and the new crop grew slowly. Tomatoes that typically ripen 90 days after being transplanted into the fields were struggling to produce a crop after 110 or 120 days.

When the tomatoes did finally ripen, they landed in a market that was awash with overproduction from Mexican fields and Canadian greenhouses as well as the surge of replanted Florida tomatoes. Prices to handlers dropped as low as $3.50 for a twenty-five-pound box, less than it cost to pick and pack them. It wasn't even worth harvesting the fields. Tens of millions of dollars of tomatoes were left to rot.

"It was a double-whammy," said Brown. "We got hit when we lost the crop. Growers who had invested millions of dollars got nothing in return. And once there were no longer any Florida tomatoes on the market, prices soared to over twenty dollars a box. Mexicans weren't affected by the freeze and they made a killing. Managers of quick-serve restaurants balked at the high prices and cut back on the amount of tomato-based items on their menus. They just walked away. So when our tomatoes finally ripened and the volume on the market returned to normal, we lost our shirts."

The freeze marked the second time in two years that the industry in Florida had to struggle through a disaster. In 2008 the U.S. Food and Drug Administration implicated Florida tomatoes in a massive salmonella outbreak. At the time, Florida had $40 million worth of tomatoes picked and ready for shipping. Consumers abruptly stopped buying, and fast food chains cancelled orders. It turned out to be a false alarm. Food safety inspectors determined that the outbreak originated in jalapeños from Mexico. After six weeks of investigation, the Food and Drug Administration completely exonerated Florida tomatoes, but it was too late for producers who had lost an estimated $100 million in sales.

For some of their financial difficulties, Florida farmers have only themselves to blame. The infrastructure of the biggest sector of the industry is based on a technology that dates back to a simpler time when supermarket produce sections offered one type of slicing tomato, usually sold three-in-a-row in cellophane-wrapped cartons. During the winter, they were most likely grown in Florida. Today's consumers demand variety. In the winter, my small town's Shaw's grocery store, which is a produce desert compared to larger, more urbane supermarkets, features a four-tiered display offering ten different varieties of fresh tomatoes. In addition to Florida rounds, I can buy cherry, grape, plum, on-the-vine-cluster, hydroponic, and organic tomatoes. Those tomatoes journey from greenhouses in Vermont and Canada and fields in Florida and Mexico—but mostly Mexico.

Although it has not officially been declared, a tomato war has raged for the past two decades between Florida and Mexico. Florida is losing. When it started, Mexican imports accounted for about one-fifth of the U.S. tomato consumption. That figure has since risen to one-third. The North American Free Trade Agreement, which came into force in 1994, certainly helped Mexico. Almost immediately after the treaty was enacted, American growers claimed that Mexico was dumping tomatoes on the market at prices below what they cost to produce. Rather than risk tariffs or other sanctions, Mexico agreed to a settlement that established a minimum value at which it would offer tomatoes for sale to this country. But Mexico still had several key advantages over Florida, among them better weather, better soils, and lower wages. But the Mexicans' far-sighted business strategies also played a role. In the early 1990s, Mexican growers pioneered new production techniques. Instead of producing "gassed green" tomatoes, they opted to plant newly developed "extended shelf life" varieties bred by Israeli horticulturists that could be allowed to ripen on the vine and still survive shipment to distant markets, depriving Florida growers of their geographical advantage. The Mexicans also adopted greenhouse culture, which has helped increase their share of the fresh tomato market. The profits reaped during the 2010 freeze left Mexican farmers flush with capital to invest in expansion at a time when Florida growers were just hoping to hang on long enough to plant another season's crops. "They made enough during that freeze to keep their foot on our neck for a decade," Brown said.

While it is true that Florida's tomato production is dominated by large agribusinesses, they are mostly family held, private companies that lack the financial leverage of most corporations. They are tiny compared with their fast food, supermarket, and institutional food-service customers. They are also dwarfed by their suppliers, who are multinational corporations such as Monsanto, DuPont, and Bayer CropScience. Almost everything a tomato farmer buys to raise a crop is petroleum based—chemical fertilizers, pesticides, plastic row covers, plastic bins, and fuel for tractors and trucks—and prices rise in lockstep with a barrel of oil. Little wonder that bankers are none too eager to lend money against a future harvest. Owners have to dig into their own bank accounts to get through lean years. And an increasing number are no longer willing or able to do that. Twenty years ago, there were about three hundred commercial tomato farms in the state. Currently there are fewer than seventy-five, and the number continues to shrink. "If Americans want imported food, they'll be dependent on imported food before they know it, because we'll be broke and gone," said Brown.

Florida tomatoes also face pressure from greenhouse and hydroponic producers located in Canada and the United States as well as Mexico. "The greenhouse market has just exploded in the last decade. There have been fascinating volume shifts," said Brown, who holds their products in the same disdain that foodies reserve for Florida's mature greens. "Our tomatoes are not a manufactured product, as opposed to what is grown in the greenhouse industry. They have a standard set of plant genetics, a standard set of environmental conditions, and they squirt them out like widgets. That's why the entire retail shelf is totally dominated by greenhouse tomatoes. The consumer perceives them to be of great value because they are beautiful and on the vine and they smell like tomatoes, but that's just a gimmick. If all a tomato has is water, what's it going to taste like? Maybe we should get scratch-and-sniff stickers for our field-grown tomatoes."

Regardless of what they do to burnish their image, Florida tomato growers just can't get no respect, in Brown's view. "It's frustrating," he said. "The Coalition of Immokalee Workers needs a bogeyman. We're it. And once an accusation is hurled in the media, it never goes away and rarely gets fact-checked. It just gets repeated, over and over again. When people are convinced that we're the monsters that we've been painted to be, you don't change their minds. Doing good things and being good citizens and business people does not make the papers. We've tried to tell our story, but reporters are not interested. Yelling 'Fire!' sells."

Brown said that Florida farmers got almost no press coverage when they endowed the Farmworker Community Support Foundation, which in 2010 donated $160,000 for improved health care and early childhood education programs in South Florida. Growers have been longtime supporters of the Redlands Christian Migrant Association, which operates daycare centers, preschools, and charter schools throughout rural Florida. They funded an AIDS education program for male migrant workers. They backed a campaign to provide dental care for pregnant farmworker mothers. Tomato money provides scholarships for Immokalee high school graduates to continue their education. Exchange members have underwritten a scholarship at the University of South Florida's College of Education. In the 2008–2009 crop year, the Tomato Committee donated nearly $300,000 to fund university research projects.

The industry was also a pioneer in food safety, Brown said. "When we voluntarily started down that road, all my friends in the produce association business said that we were nuts. 'Just stonewall the regulators,' they said. We said that the right thing to do was figure out how we could do things better to try to lessen the chances of disease outbreaks caused by tomatoes. We worked with the Food and Drug Administration, and we came to the party long before the leafy-greens people came. We wanted to do whatever we could to prevent a major foodborne-illness crisis."

To uncover labor abuses in the fields, the Florida Fruit and Vegetable Association established a group called Socially Accountable Farm Employers (SAFE) in 2005. Reggie Brown is on the board of directors and is listed as the organization's contact person. The non-profit organization was established to provide independent auditing to make sure that certified farms used fair and legal employment practices and to make sure the fields were "free from hazard and violence." These were some of the steps the coalition's Campaign for Fair Food was demanding farmers take, but from the outset there were complaints that SAFE was a case of the industry fox guarding the henhouse if there ever was one. No one representing Florida migrant workers sat on the SAFE board, although three of the five members headed organizations that had received generous financial support from tomato growers and other farmers. The skeptics' position was vindicated in 2007, when the SAFE auditor declared that the fields of Immokalee were slavery free only days before the high-profile Navarrete case came to light, and further vindicated a year later when court proceedings revealed that Navarrete slaves had worked on farms controlled by two SAFE-certified companies.

Brown insisted that Florida tomato farmers abhorred slavery as much as anybody. "But there's slavery in other places, too, in the United States. You go to any city where you have nail parlors or Chinese restaurants, and you're going to be able to find human trafficking." As for the housing conditions in Immokalee, he agreed they were an embarrassment to the industry but pointed out that the farmers did not own the decrepit Immokalee trailers. In other areas where they did provide housing for their workers, he noted, the accommodations were government regulated and inspected by the health department.

Tomato growers, he claimed, complied with the same labor laws as other employers. "We pay the same wage that McDonald's and Burger King pay in their shops to the people that work the counter. It's minimum wage. That's the law of the land. But because we have a seasonal business, our employees may not work for us twelve months a year, but in the period they are working for us, they're making minimum wage." People who accuse the tomato growers of not paying the workers enough to live off of, he said, are only looking at one slice of a migrant's income stream. "That worker might be employed by eight or ten different companies during the course of twelve months. He could be in North Carolina in the summer picking tomatoes and New York State picking apples in the fall before he comes back here."

The proof that growers pay competitive wages, Brown contended, is that there is no labor shortage in the tomato industry. Their rates, he maintained, match those paid by other employers for jobs requiring similar skills and abilities. "Otherwise, they'd work for someone else. And some of these folks have been working for us for five or ten years. It's hard work, but it's good work."

Brown was given an opportunity to present his industry's side of the story to the U.S. Senate Committee on Health, Education, Labor, & Pensions during an April 2008 hearing on improving working conditions for tomato workers. "It was the worst day in my entire career up to that point, and the toughest," he said. "You can imagine yourself trying to make an honest presentation of how we saw the issues, when the rest of the table knew for sure that I was the devil incarnate, including the senators, all Democrats. There was not a friend in that hearing room. It was no good to be falsely accused and so defamed as an industry for something that we weren't doing. But Senate hearings are an art—almost like bull baiting."

Senator Bernie Sanders (I-VT) presided over the hearing. Senators Edward Kennedy (D-MA), Richard Durbin (D-IL), and Sherrod Brown (D-OH) were also present. All of them had reputations for being vehemently prolabor. In their opening statements, the senators focused on wages, honing in on two claims that the Tomato Exchange had made. The first, which appeared on its Web site and was later repeated by Reggie Brown at the hearing, was that the wages paid to tomato pickers averaged over twelve dollars an hour. Senator Durbin asked that the committee join him in doing the math. At the going rate in 2008 of forty to sixty cents per thirty-two-pound bucket, a harvester would have to pick about three thousand tomatoes each hour—nearly one per second. He would have to fill a bushel-size bucket, run over to the truck, dump it, and run back to his assigned row every two minutes. "Is that physically possible?" asked Durbin. "I don't think it is."

In his testimony, Lucas Benitez of the coalition called Brown's bluff. He said that he could refute the claim that pickers could get twelve dollars an hour by citing reports from the U.S. Labor Department and respected sources within the produce industry. "I want to make this issue as clear as possible. If Mr. Brown can guarantee that $12.46 an hour, backed up by a verifiable system of hours with time clocks in the fields and thereby eliminate the antiquated system of work by the piece, then we will take it. However, unfortunately, I don't think that I have to be a fortune teller to know what the response will be."

The hearing moved on to the exchange's refusal to pass the penny-per-pound rate increase along to the workers. Brown said that in theory the penny-per-pound idea seemed like a good one. "For the record, we do not object to the fast food chains paying extra to the workers who pick the tomatoes they buy. Our members simply do not want to be part of that arrangement."

One of the difficulties, he said, was that it would be impossible to determine which pickers harvested the tomatoes destined for the chains who had agreed to the Campaign for Fair Food's requests.

"During harvest, tomatoes that the workers pick are not individually identified or labeled by worker or by customer. At the time of harvest, a tomato picked by a worker could ultimately be purchased by any number of the producer's direct customers," he said. And those customers would in turn buy tomatoes from several producers and intermingle them. Brown said that growers feared they would be open to lawsuits from workers who were treated unfairly. "Workers also could allege that there is/was a scheme to defraud them and each check issued [allegedly in an incorrect amount] could be a separate bank or wire fraud. This is by definition a RICO [Racketeer Influenced and Corrupt Organizations Act] case. What's more, RICO allows plaintiffs to bring additional grounds to allege fraud-based activities on whatever size enterprise they seek to attack." The exchange's members, said Brown, were also concerned that the extra-penny-per-pound program constituted an attempt to restrain trade.

Sanders asked Brown if the legal opinions were his idea, or whether he had gotten opinions from attorneys.

Brown replied, "We purchased legal opinions from legal firms in this country to affirm those opinions, yes."

Sanders countered that the fast food companies who entered into the penny-per-pound agreement had also sought legal advice. "Yum! Brands—and this is, as you know, not a small company. This is a huge corporation that, I gather, has the money to hire expert legal advice," said Sanders, pointing out that the committee had received a letter from the fast food giant's senior vice president saying, "Yum! Brands's attorneys are fully confident that the agreement is legal." Sanders also reported that the committee had received a letter signed by twenty-six law professors from around the country stating, "The ostensible legal concerns of the growers exchange are utterly without merit. Growers who comply with the agreements will not violate antitrust, labor, or racketeering laws. The unfounded assertions of the growers exchange should not deter any grower from adhering to the agreements, nor should those assertions deter any fast-food company or other buyer from entering into similar future agreements. The only real . . ." Here Sanders paused for effect and said, "I would like you to listen to this. This is according to twenty-six law professors around the country." He resumed, "The only real antitrust concern would arise if several growers agree among themselves to not participate in the monitoring program." Sanders also pointed out that lawyers representing McDonald's had issued opinions similar to that of the law professors. "I gather McDonald's has the resources to hire some pretty good lawyers," he said. Finally, he told Brown that the Senate committee itself had engaged two nationally recognized law firms to look at the Campaign for Fair Food agreement and that they had concluded that any claims that the terms of the agreement violated the Sherman antitrust law appeared meritless.

Brown replied, "Senator, that is one group of legal opinions. Our legal opinion is different, okay?"

"You might want to reconsider the attorneys that you are currently consulting," said Senator Sanders.

"I was fried," Brown told me.

That seemed like an accurate summation of Brown's bad day in the Dirksen Senate Office Building. But, although Tomato Exchange removed the threat of $100,000 fines to members who cooperated with the coalition, members of the group otherwise remained defiant. Two and a half more years would go by before the Tomato Exchange capitulated.

In early 2009, after the sentencing of the Navarrete family members for slavery, then Florida governor Charlie Crist agreed to meet with coalition members. It represented a landmark. The workers had been asking Florida governors to sit down with them and discuss their plight for fifteen years—through four administrations, both Democratic and Republican—and Crist was the first to do so, perhaps inspired by the effect that national media reports on working conditions in his state might have on his impending (and ultimately unsuccessful) run for the U.S. Senate. After the meeting, Crist wrote in a letter to Benitez and Reyes:

> I have no tolerance of slavery in any form, and I am committed to elimination of this injustice anywhere in Florida. I unconditionally support the humane and civilized treatment of all employees, including those who work in Florida's agricultural industry. Any type of abuse in the workplace is unacceptable.
>
> I support the Coalition's Campaign for Fair Food, whereby corporate purchasers of tomatoes have agreed to contribute monies for the benefit of the tomato fieldworkers. I commend these purchasers for their participation, and I encourage the Florida Tomato Growers Exchange and its members to participate in the campaign so that these monies can reach and provide assistance to the workers.

Florida's agricultural industry has always had close relations with the state's politicians. Crist's letter must have stung, but it didn't alter the exchange's position.

Although major grocery chains (with the notable exception of Whole Foods) still held out, the Campaign for Fair Food had so far been a phenomenal success. Together, the fast food and food-service companies that had signed agreements bought hundreds of millions of dollars worth of Florida tomatoes each year. Their market share represented a huge financial incentive to any large tomato grower. If bad publicity was the coalition's stick to prod recalcitrant producers, the buying power wielded by its roster of corporate partners was its carrot, and the carrot was growing bigger and juicier with each passing season. It seemed like only a matter of time before a large farmer broke ranks with his peers to get his share (or more) of that business. In the fall of 2009, that break occurred. East Coast Growers and Packers quit the exchange and signed on with the coalition. In a lesson that was not lost on other Florida producers, Chipotle Mexican

Grill, the burrito chain with more than nine hundred restaurants across the country, immediately announced that henceforth it would be getting its tomatoes from East Coast Growers and Packers.

If supporters of the coalition thought that East Coast's defection would have an immediate domino effect, they were wrong. When I met with Brown, the growers were still hanging tough. I asked if the exchange was in talks with the coalition, and he said, "We don't have any contact with the Coalition of Immokalee Workers. We don't have anything to do with them." He said that he really didn't know what the coalition was. Was it an unofficial union? Was it a community organization? Was it a grassroots movement?

Five months later almost to the day, an unexpected telephone call came into the coalition's office. Reggie Brown was on the line. The exchange was ready to join the Campaign for Fair Food.

In the intervening months, two other major tomato operations agreed to the coalition's terms. Heading into the 2010–2011 season, after having been battered the previous year by the catastrophic freeze and subsequent collapse in tomato prices, and still stinging from the negative attention brought on them when it was revealed that the Navarretes' enslaved crew had worked on their farms, Pacific Tomato Growers and Six L's came aboard. With three of their largest competitors now part of the campaign, it was no longer feasible for the exchange's other members to hold out.

A little after one o'clock in the afternoon on November 16, 2010, Brown sat down beside coalition members Reyes and Benitez at a folding table set up in the backyard of the coalition's headquarters for a press conference.

"Our industry is and always has been strongly committed to supporting real, long-term and comprehensive solutions that improve the lives of all farm workers and their families," he read from a statement issued by participating farms. "That's why we have agreed to work with the CIW in establishing new standards of verifiable social accountability for the tomato industry as a whole. We realize that this is a work in progress and that this partnership will get stronger over time. It will not be completed overnight. As time goes by, we are confident that we will be able to weed out the bad actors and, working together, build a stronger, more sustainable industry that will be better equipped than ever to thrive in an increasingly competitive market place."

No other fruit or vegetable growers' group in the United States had ever inked such a far-reaching deal with representatives of its workers. With the stroke of a pen, Florida's tomato producers had put themselves on the road to becoming the most progressive employers in the country's produce industry. Under the terms of the new arrangement, the 2010–2011 crop year would involve only Pacific and Six L's. These two companies would work with the coalition to iron out practical issues such as the ones Brown had raised in the Senate hearing and build a template for full implementation of the Fair Food Code of Conduct, which would be rolled out to all participating growers the following year.

With a major victory behind it, the coalition refocused its efforts on the supermarket chains. By securing the exchange's cooperation, a conduit had been opened for those extra pennies per pound. The onus had shifted to the grocery stores to join the fast food industry and others and pay a little more for their tomatoes. "Make no mistake," Benitez cautioned at the press conference. "There is still much to be done. This is the beginning, not the end, of a very long journey."

Sitting together at that table in the coalition's backyard, Reggie Brown and Lucas Benitez could certainly agree on that.

MATTERS OF TASTE

In early 2010, I enjoyed a supermarket experience that I'd never had before. I bought a pretty, stridently red winter tomato that actually tasted like something. It was by no means a great tomato, harboring only hints of the flavor wattage of a vine-ripened August tomato, but it was nonetheless unmistakably a tomato, in taste as well as appearance. As is de rigueur with so-called premium produce nowadays, my purchase, which weighed three-quarters of a pound and cost $3.47, versus 80 cents a pound for its nearby commodity cousins, bore a little sticker with the trademarked name Tasti-Lee. A few days later, I returned to the same store hoping to replenish my tomato larder, but the Tasti-Lees had all been sold. And I was left to ask, what made this tomato so different? How come I had never heard of it? Why don't all supermarket tomatoes taste like it did? And where could I get more?

With those questions on my mind, I drove to the University of Florida's Gulf Coast Research and Education Center, a curvilinear structure faced with pinkish bricks that rises like a space station from the endless fields of strawberries and tomatoes a half hour's drive southeast of Tampa. There, in an office crammed with all manner of tomato kitsch—coffee mugs, antique labels for packing boxes, framed vintage magazine advertisements for Campbell's Tomato Soup, tomato piggy banks, tomato salt and pepper shakers, and teetering stacks of—who knew—*The Tomato* magazine—I met John Warner Scott, a professor of horticultural sciences. Scott, who is known far and wide in the tomato business as "Jay," is one of the most prolific breeders of new tomatoes in the state. Over his three-decade career at the university, he has developed more than thirty varieties, although he doesn't keep track. "I haven't gone back and counted in a while," he told me. Unlike seedsmen who rely on molecular biology, DNA sequencing, and in the case of some crops, genetic modification, Scott is the last of a dying line of old-fashioned plant breeders. His tools are the same ones the great nineteenth-century tomato grower Alexander Livingston used to develop the Paragon: a keen eye, a disciplined palate, and superhuman reserves of patience. Each year, Scott grows several hundred different varieties of tomatoes, called "parent lines," in test plots surrounding the Gulf Coast center. His goal is to find plants with complementary traits—one may have disease resistance but low yields, another high yields but weak immunity—and crossbreed them hoping that some of the offspring will carry the best traits of both parents. Toiling in the hot sun, Scott pulled a floppy sun hat over his close-cropped graying hair and plodded through the rows, notebook in hand, carefully examining each plant and ticking off a mental checklist: How many fruits has it set? Are they big? Do they have cracks? Are their bottoms smooth and rounded, or do they still have scar tissue where the blossoms fell off? What's the color like? "With some of them, you can just look at the plant and just throw it out," he said.

If a plant passed visual muster, Scott took out his pocket knife and, still standing in the field, lopped off a slice and tasted it. "Plant breeding is a matter of seeing what's good," he said. "But you can't make any decisions based on one season. You have to grow a variety a lot of times in a lot of environments to see if it's really good."

Tasti-Lee is not perfect. Its fruits are smaller than commercial growers like. But it is as close as Scott has ever come to finding Tomatoland's Holy Grail—a fruit thick-skinned enough to shrug off the insults of modern agribusiness, but still tender at heart and tasting like a tomato should.

It's been an enduring challenge for tomato breeders. Modern tomatoes consistently rank at or close to the bottom of consumer satisfaction surveys. In the vivid words of one writer:

> As for the reasonably fresh tomatoes that one used to be able to buy at fruit-and-vegetable stores in the city, they seem to have all but disappeared, even during the local tomato-growing season. In their place is something that is called a tomato, that has the shape of a tomato and a tinge of the color of a tomato, and that sells at fancy-tomato prices, but serious doubts have been raised about whether it tastes like a tomato. Such objects can be had readily at any supermarket in the depths of winter or in the heat of summer. They lie in pale pink piles or in narrow plastic cartons, three or four to a box. Some hard words have been uttered and written about what has happened to the American tomato in recent years. "Insipid," "blah," "tough," "like eating cardboard," and "plastic junk" are terms that people apply to store-bought tomatoes today. Craig Claiborne, of the *Times*, has called the tomatoes commonly on sale "tasteless, hideous, and repulsive."

Those words were written by *New Yorker* writer Thomas Whiteside in 1977. Nearly thirty-five years later, Scott and a handful of other breeders are saying that it is high time someone fixed the taste problem.

Developing a better tomato can take years, and even then, there is no guarantee that it will be picked up by professional growers and have a shot at commercial acceptance. Florida's multimillion-dollar tomato industry is littered with once promising but now forgotten varieties. But in early 2010, after more than a decade of painstaking growing, breeding, and crossbreeding, Tasti-Lee left the rarified confines of academic test plots and rigorously monitored consumer-tasting panels to try to make its way in the competitive hurly-burly of the produce section. If Tasti-Lee lives up to its early promise, Scott will achieve a plant breeder's version of immortality. The rest of us finally will be able to head to the local supermarket any day of the year and bring home a half-decent-tasting tomato.

But it won't be an inexpensive tomato. Scott developed the Tasti-Lee to provide farmers in his state with a crop that can be planted outside to compete with hydroponic, greenhouse-grown tomatoes, the latest competitive threat to the Florida fresh tomato industry. Beginning from almost nothing in the early 1990s, greenhouse tomatoes expanded from a tiny niche-market novelty mostly imported from Europe to a mainstream produce item. They are now in every supermarket and account for about 10 percent of fresh tomato sales. Although Florida's field-grown slicing tomatoes remain as popular as ever in the food-service industry, sales have declined sharply in supermarkets. With Tasti-Lee, Scott hopes to give growers a baseball-size tomato that packs the same flavor as the popular ping pong ball–size salad tomatoes produced in greenhouses and often sold in clusters on the vine. "It seems to me that it would be a win-win situation," said Scott. "Consumers tend to be spoiled. They go into the grocery store and they expect to see fresh tomatoes any time of year, even if they grumble about the quality. I want people to buy Tasti-Lees because they like them, not just because they are the only tomatoes there."

Like many plant varieties, Tasti-Lee owes its existence to a combination of serendipity and the time-sharpened instincts of a great plant breeder. In Florida, the summer of 1998 was a terrible season for anyone trying to grow a tasty tomato. For some unknown reason—too wet, too cloudy, too hot—Scott's tomato field tests failed to produce fruits with any sweetness. Even tried-and-true varieties that had been sweet during previous years tasted dull. But one morning after tasting fifty varieties, each more bland than the other, Scott spotted a nice-looking tomato called Florida 7907. He picked a fruit, cut off a wedge, and popped it into his mouth. "Aha!" he said.

It was sweet, but Florida 7907 had one big flaw that made the variety a nonstarter for commercial production: It was too spherical. Florida growers like their fruits to have defined shoulders and slightly flattened bottoms. And that's only one item on a list of must-haves. Because producers are paid strictly by the pound, plants first and foremost must produce high yields of large, uniform fruit. They have to be able to resist diseases and tolerate extremes of heat and cold. And their tomatoes need to have a long shelf life. Taste enters the equation, if it enters at all, only after all those conditions are met. "Sometimes I wonder why we even bother with flavor," said Scott. "There is no easy way to breed for taste. It's not like there's one genetic marker that tomatoes must have to taste good," he said.

The structure of a tomato also makes breeding for both taste and toughness a difficult balancing act. The gooey part of a tomato, called locular jelly, has most of the all-important acidity. The pericarp tissue, the walls of a tomato, give it strength and some sweetness, but no acidity. The harder a tomato is, the more bland it is likely to taste. Even if you have a perfect balance of sugars and acids, there are still many obstacles in getting decent-tasting tomatoes from field to consumers' kitchens. Most Florida tomatoes are picked at the so-called mature green stage. Under ideal circumstances, a mature green tomato, reddened by being exposed to ethylene gas, will ripen and develop a measure of taste—not great taste, but something. The problem is that short of cutting one open, there is no definite way to tell a mature green tomato from one that is simply green. Inevitably, some immature tomatoes get picked, and they will never develop flavor, although the ethylene will give them the appearance of ripeness. Finally, even if all else goes according to plan, a tomato can lose its taste if exposed to cold temperatures at any time between harvest and being eaten, after which point it can never recover it.

Crop specialists even have a scientific term for this process: "chilling injury." Whether it happens in a truck, warehouse, produce section, or home refrigerator, a tomato that is held at temperatures lower than 50 degrees soon becomes a tasteless tomato. For reasons unknown, chilling reduces the fragrant volatile chemicals that are all-important in giving the fruit its distinctive flavor. Unfortunately, keeping tomatoes cool extends their shelf life, too, so the temptation to refrigerate dogs tomatoes every step of their journey to the table. Years of efforts by a plant breeder can be destroyed by a few days in a refrigerator.

Scott was also developing a line of what he calls "ultrafirm" tomatoes during the same season he happened on the sweet-flavored 7907. Among those he was developing was a tomato called Florida 8059. It was hard and had the right shape. Sensing a match made in heaven, Scott crossbred the sweet but too-spherical 7907 with the firmer 8059, and in the fall of 2002 the first of what was then referred to as Florida 8153 ripened. Scott thought the new hybrid carried the best traits of both parents. At trials conducted by the university, consumers on test panels agreed. Time after time, 8153 beat out other tomatoes. Subsequent chemical analyses showed that the fruit had a desirable balance of sugars, acids, and volatiles. It also had a surprise bonus: Both of its parents possessed what plant breeders call the "crimson" gene, which was originally revealed when the pioneering tomato geneticist Charlie Rick crossed a wild *L. chilense* (a relative of the domestic tomato) with a commonly grown variety. The crimson gene gives 8153 a striking fire-engine red color and an extraordinarily high level of lycopene, a sought-after antioxidant. "It sounds like magic, doesn't it?" said Scott. "It really is, in a way."

Florida 8153 had everything going for it, except for a catchy, appetizing name. Scott christened and trademarked his new baby Tasti-Lee, Lee being the first name of his mother-in-law, a tomato lover who had encouraged and supported his research through the years. "You hear lots of stories about bad mothers-in-law, I had a great mother-in-law," Scott said, a flash of emotion overcoming his usual deadpan. "She had tasted what was then still just called Florida 8153. She really liked it and encouraged me. Sadly, she fell terminally ill. I went to visit her in the hospital. She was in a coma at that point, but I took in a tomato anyway and showed it to her and told her that I was going to name it after her. I like to think she heard me."

Four seed companies lined up to bid for rights from the university to produce and distribute Tasti-Lee seeds. The winner was Bejo Seeds, Inc. A large, family-owned, Dutch firm with offices around the world, Bejo's specialties are cabbage, carrots, and other cool-weather crops. "We felt that marketing would be a key to Tasti-Lee's success," said Scott. "It seemed like Bejo would be hungry to get into the tomato market and that they would push Tasti-Lee pretty hard."

The job of giving Tasti-Lee that push fell to Greg Styers, Bejo's sales and product development manager for the southeastern United States, who has been known to board airplanes lugging twenty-five-pound boxes of tomatoes as carry-on baggage. "We had a vision to start with a grassroots movement," said Styers. "We were going to start with roadside growers and chefs. People who were interested in good flavor and good quality. Then we were going to work our way up." It didn't turn out as planned. Styers, who was looking for a grower who shared his vision that Tasti-Lee was "born to be a premium tomato," approached Whitworth Farms, which grows vegetables on seven hundred acres near Boca Raton, making it a small player in the Florida tomato business. "Whitworth was big enough to deal with some large retailers, but small enough that they were willing to take a chance on Tasti-Lee. It was a perfect fit for us," said Styers.

One of Whitworth's customers was Whole Foods Market. Glenn Whitworth, who owns the farm along with his sister and two brothers, approached one of the company's produce buyers. Weeks went by before the buyer would even schedule a meeting with Styers and Whitworth. When they did finally get some time, Styers stopped by a Whole Foods store beforehand and bought one of every tomato on display and added a Tasti-Lee to the mix. On the basis of that impromptu conference room taste test, the buyer agreed to test-market Tasti-Lee. In February 2010, Tasti-Lees began appearing in sixteen Whole Foods stores in Florida. By late March, reorders were coming in faster than Whitworth could grow Tasti-Lees. Later that spring, Whole Foods stores as far north as Washington, DC, began to carry Tasti-Lees, and by the end of the year, other retailers and even a few restaurant chains were expressing interest. "I think the stars really lined up for Jay when he developed this variety. It truly is remarkable," Styers said.

Scott, who drawls his carefully chosen words with little inflection and almost no emotion, didn't go that far. "I stand behind it," he said. "For a full-size tomato, it's better in my opinion than what's out there. Hopefully, it goes." If it doesn't, Scott has plenty to keep him busy. He's currently developing heat-tolerant tomatoes, tomatoes with resistance to the virulent leaf-curl virus, and tomatoes that can be grown on the ground and theoretically harvested by machine. And he hasn't given up on flavor. "In some work we've done, there is this fruity-floral note that adds pique to the sweetness," he said. "We've crossed a big, crimson tomato with that trait into one of Tasti-Lee's parents. The result might have even better flavor."

In the race to build a better tomato, Scott faces stiff collegial competition from Harry Klee, a fellow University of Florida professor who works out of a laboratory in Gainesville, a few hours' drive north of Scott's test plots. Although the goal is the same—a tomato that can be grown commercially in Florida and come out with taste intact—the two researchers are taking diametrically opposite routes to get there. And they bring differing personalities to the problem. Klee, fifty-seven, is tall and athletic and looks like he's ready at any moment to push himself away from the computer keyboard for a quick round of pick-up basketball. He speaks in eloquent, neatly structured sentences, posing Socratic questions, as he might during a classroom lecture, then answering them himself. He zips around campus in a nonprofessorial two-seater BMW convertible and brings the same unorthodox approach to the tomato project. Instead of focusing on the types of tomatoes that have dominated the market—ones favored by growers and shippers, who want toughness, disease resistance, and bulk—Klee started with consumers, who are crying for decent-tasting fruit. "We have two goals," he told me. "One is to define what a good-tasting tomato is, and two is to find the genes that control the processes that make good taste and breed them back into tomatoes. To do that, we are bringing together molecular biology and psychology."

When I reached him by telephone to set up a time to discuss his work, Klee immediately enlisted me as a guinea pig. That's how I found myself one sunny February morning sequestered in the sensory testing lab of the University of Florida's Department of Food Science & Human Nutrition. But before I was allowed to put a morsel of tomato in my mouth, I had to undergo a battery of psychological tests to determine how much of a foodie I am. I sat and filled out a form that asked questions that had nothing to do with tomatoes and at times became too personal for comfort. What was the strongest dislike of any kind that I had experienced? The name of a former boss leapt to mind. Strongest liking? My three daughters, of course. The strongest sensation I had ever experienced? Kidney stones. I was asked what the most pleasant experience I had ever had was, where I would rank my most pleasant taste experience compared to that, and where the taste of the best tomato I'd ever eaten stood in comparison to the best thing I'd ever eaten. The goal of these questions was to determine whether taste was an important sense to me, and if so, where tomatoes ranked among my preferences, or whether I even liked them at all. After I handed my survey to a graduate student, a sliding door opened in a plain white wall in front of me, and a pair of hands in surgical gloves pushed through a plate containing a halved cherry tomato. I tasted. It lacked the acidic balance of a truly great tomato. I gave it a score of 60 on a range of -100 (ghastly) to 100 (exquisite) on the keyboard in front of me and waited for another offering. In all,

I chewed through six varieties of tomatoes: Red Pear, Cherry Roma, Ailsa Craig, Matt's Wild Cherry, Tommy Toe heirlooms, and, as a control, a Cherry Berry hybrid bought at a local Wal-Mart.

In his efforts to build a better tomato, Klee has joined forces with a multidisciplinary team that includes psychologists, food scientists, statisticians, and molecular biologists. "It is a very achievable goal," said Klee, a professor in the Horticultural Sciences Department. "I'll predict that within five to ten years, you'll see significant improvements in the flavors of industrial tomatoes. The seed companies have finally woken up and realized that there is a big problem with lack of flavor and that people are willing to pay for better-tasting tomatoes. They see that there is money to be made there." An alumni of the genetic engineering giant Monsanto, Klee allows that the process would be quicker and simpler if he were to use genetic modification—simply taking the desirable genes from a tasty tomato and splicing them into the DNA of industrial fruit. That is precisely the sort of work he did at Monsanto, where he designed a tomato that had a much longer shelf life than conventional fruits. "It worked beautifully," he said. "But we developed it at the time when the public turned against genetically modified foods, so Monsanto dropped the whole program." Recognizing that consumers are wary of genetically modified plants, Klee has chosen to go the slow, painstaking route this time—one of the advantages of being in academia, he said. Much like Scott, Klee uses traditional breeding techniques of cross-pollinating plants and sorting through thousands of their offspring, hoping to find one that has the traits they seek.

Klee is convinced that tomato breeders took a wrong turn fifty years ago. "If you ask commercial seed companies why they are making tomato varieties that have lost all their flavor, the answer is very simple," he explained. "They have focused all their energies on their customers. Who are their customers? The commercial growers. What does a grower get paid for? Yield, size, and appearance. They make more money for very large tomatoes than they do for small ones. The grower is not paid for flavor. So you have a fundamental disconnect between what growers want and what consumers expect."

By these criteria, no one can say that breeders have failed to give factory tomato farming the plants it wants. Per acre yields of tomatoes have gone up an astounding fivefold since the 1930s. Since 1970 alone, they have increased by a factor of three. "It's incredible," said Klee. "If you think of any other agricultural crop where you've had increases like that, it becomes mind boggling. If you did that with corn, you'd be feeding the world. But as you focus on making the tomato bigger and firmer, you are ruining the flavor, pure and simple. Yields have simply outpaced the plants' abilities to fill the fruit with flavor and nutrients. What we have ended up with is something that's large but has basically had all the good points diluted out of it. They've essentially taken the package and added water. Strawberries are the same story, but tomatoes are probably the worst example."

Neglected for a half-century, the genes that once gave commercial tomatoes taste have become lost. To rediscover those genes, tasting panels such as the one I joined are working their way through 150 varieties of nonhybrid heirloom tomatoes, survivors from a time when taste mattered to growers. However, identifying what makes a tasty tomato is anything but straightforward. Although the statement would be sacrilege to any food snob worth his or her Himalayan salt, Klee asserts that not all heirlooms taste good. To prove his point, he handed me a golf ball–size dark red fruit. "This is a Stupice heirloom," he said. I bit in and winced. It tasted musky and sour—worse than any store-bought fruit, whose main sin is utter lack of taste, not taste you can't wait to rinse from your mouth.

"Tomato flavor is really complicated," Klee explained. "And because of that complexity, not much science had been done on it until we started ten years ago." A combination of sugars, acids, and volatiles (the technical name for chemicals we can smell, often at minute levels measured in a few parts per billion) determines the tomato taste. Having the right balance of sugar and acids—mainly in the form of citric and malic (the latter is responsible for the tart taste of green apples), with more citric than malic—provides a foundation on which tomato taste can be built. But since most of what we perceive as flavor is actually aroma, it is the fifteen or twenty volatile compounds that have the biggest impact on tomato taste. They have names that you are not likely to find on the menu of your favorite restaurant: cis-3-hexenal, beta-ionone, beta-damascenone, 1-penten-3-one, 2+3 methylbutanal, 2-isobutylthiazole, 1-nitro-2, methyl salicylate, and phenylacetaldehyde. Of those, perhaps a half-dozen are critical. Without them, a tomato will not taste like a tomato. "You've got all these different compounds that are all synthesized by different chemicals independent of each other, so you have a huge scientific problem to solve," Klee said. "We've identified fifty genes that affect flavor." With other fruits, the chemical equation is much simpler. A banana, for instance, owes its distinctive flavor to a single chemical, isoamyl acetate. Furaneol is closely linked to strawberry taste.

Klee took the cap off a vial containing a clear liquid and waved it under my nose. I got a dizzying snootful of Juicy Fruit gum. "You're smelling beta-ionone," he said, passing me another vial. I sniffed again. There was no mistaking the summery fragrance of roses. "That is 2-phenylethanol, and it is actually a major component of rose scent," he said. The next vial brought my winter-weary nose profound memories of fresh-cut grass after spring's first mowing. Klee said it was cis-3-hexenol. A vial containing beta-damascenone had woodsy and fruit flavored notes that I associated with grapes and wine. Juicy Fruit, roses, cut grass, grapes—none of these volatiles smelled anything remotely like a tomato, yet Klee believes that all of them have to be present to deliver the fruit's signature flavor. Klee explained, "You need the whole package. If you bit into a tomato that was really high in 2-phenylethanol, you'd say, 'That tastes like a rose.' There is no one chemical that you'd smell and say, 'Oh, tomato.' It's a combination of all of them."

The taste panel was part of Klee's attempt to identify which volatiles in what concentrations make a tomato taste good. The small heirloom called Cherry Roma—the epitome of the tomato's dance between sweetness and tartness—has consistently won top marks. Larger varieties such as Bloody Butcher and Brandywine, much beloved by home gardeners, have also scored well. Once the test panels had identified about twenty varieties that consistently scored highly, he chopped those tomatoes up and placed his scientific salsa in a machine called a "gas chromatograph"—in essence, an artificial nose—to "sniff" out volatiles in the choice breeds.

In consultation with Howard Moskowitz, a renowned food scientist who has worked with major companies like Pepsi, Campbell's Soup, and General Mills foods to develop new products (his discovery that some consumers like chunky spaghetti sauce made Prego a runaway success and gave birth to an entire class of bottled spaghetti sauces), the University of Florida researchers devised a computer model to provide scientific underpinning to the preferences of the hundreds of participants in taste panels. Klee unfolded a printout that looked like an electrocardiogram, with wavy lines running across the page. One axis of the graph listed chemicals, the other a single tomato variety. The lines represented the content of each chemical in the tomatoes. Klee jabbed his pen at a peak. "This beta-ionone," he said. "We're finding that tomatoes that rate highly overall always have high concentrations of beta-ionone." Using his elaborate statistical tools, Moskowitz will be able to create a "formula" for a good tomato, telling Klee the concentrations of volatiles and other chemicals he should aim for. "What we end up with is a blueprint," Klee said. "Then we have to figure out how to reproduce that blueprint."

That involves searching for the specific genes that cause tomatoes to produce beta-ionone and each of the other desirable chemicals. That search is rooted in a greenhouse a few hundred yards from Klee's lab, where he pampers some of the vilest-tasting tomatoes on the planet. They don't even look like tomatoes: They're green, and as hard, small, and unyielding as a pebble. But their beauty lies at the genetic level. By crossing wild-tomato relatives like these with domesticated varieties, botanists can see what genes produce what chemicals—a process Klee likens to discovering that a criminal you are looking for lives in California. "You've narrowed the search considerably, but you still have a long way to go."

The criminal justice analogy is apt. To zero in on the gene he wants, Klee deploys the same DNA technology that police investigators use to identify suspects. He has now discovered about half of the volatiles he thinks must be present in a good tomato. Once all of them have been found, they'll be a tool kit that breeders can use to reintroduce tasty traits into industrial-grade fruits. "There is no one perfect tomato," said Klee. "It's comparable to walking down the soft drink aisle in the supermarket. Some people are going to prefer Diet Coke, some Coke Zero, some Classic. We've found that Hispanic people prefer harder, more tart tomatoes than northeasterners. Probably because Hispanics are going to chop up a tomato and put it in salsa, while someone in the Northeast might slice it and put it in a sandwich. Different uses, different tomatoes. But there are some common traits. I think we can find them and re-create a pretty good tomato," he said.

Pretty good for consumers looking for an out-of-season tomato worth its name. But how good can a consumer feel knowing that the tasty tomato has been raised and harvested by the most abused workforce in the country?

Fortunately, a scattering of groups and individuals are trying to address that problem. They may not have PhDs in biochemistry and horticultural science (although a few have Ivy League law degrees), but they are approaching the challenges of improving the lots of the people who bring us our food with the same persistence and tenacity as Scott and Klee bring to their breeding programs. They are far from their goals, but like Scott and Klee, they know where they want to go.

BUILDING A BETTER TOMATO

THE FARMER

Road 74 bisects Charlotte County, running fifty miles due east from the Gulf Coast town of Punta Gorda toward Lake Okeechobee. It cuts a perfectly straight line through sparse, featureless fields dotted with herds of grazing cattle. Tom Beddard had provided me with the address of Lady Moon Farms, his mixed vegetable operation, but it had been twenty minutes since I last saw a street sign, house number, or for that matter, anything I would classify as a building.

My cell phone rang. It was Beddard. "Where are you now?" he asked.

"I have no idea," I said.

"You'll see our packinghouse on the north side of the road. Can't miss it," he assured me.

A few miles farther along the highway, I pulled into a sandy parking lot in front of a beige warehouselike building. The place looked deserted—no other vehicles in the lot, no workers scurrying about, no sign reassuring a visitor that this lonely outpost was, indeed, Lady

Moon Farms. I got out of my car. No one answered my knock on the front door. Around the side of the building, a lift truck stood idle beside some parked tractors and wagons. Assuming that I had stopped at the wrong place, I turned back to my car. Then I noticed a white pickup truck speeding toward the building on a lane between the rows. It stopped beside me. A middle-aged man with short salt-and-pepper hair looked me over from head to toe before unfolding himself from behind the wheel. He administered the sort of no-nonsense handshake you'd expect from a six-foot-five-inch, sun-weathered farmer. "Jump in, I'll show you around the farm."

I had come to Lady Moon to see the all but impossible, if the horticulturalists I had spoken with were to be believed. Since the late 1990s, Beddard had been growing tomatoes and other vegetable crops in South Florida using purely organic practices—no synthetic chemical fertilizers or pesticides—and succeeding on a commercial scale. Lady Moon Farms is the largest organic grower on the East Coast. Whole Foods Market is one of its major customers. In order to maintain a year-round supply of products, Beddard farms 850 acres in South Florida, 450 acres in Georgia, and 300 acres in Pennsylvania. "When I first came down here, everyone told me that you can't do organic in Florida," he said.

From where I sat in the cab, it was evident that he had proven them wrong. Square-edged rows covered tightly in white plastic stretched off to a distant cane windbreak. It was mid-October, still early in the growing season, and deep green tomato plants stood knee-high at three-foot intervals above the plastic. In the distance, a tractor crept along a row towing a sprayer behind it that filled the air with white mist. In another corner of the field, a group of Hispanic men were bent over the young plants. To me, Beddard's fields looked exactly like those of Ag-Mart, Six L's, Pacific Tomato Growers, or any other large conventional grower. But Beddard assured me that appearances were deceiving. That tractor was spraying *Bacillus thuringiensis*, a popular organic insecticide that is made from bacteria that are naturally present in the soil. The workers were wielding squeegees that had been soaked in common household vinegar to kill or slow the growth of weeds.

His biggest challenge in Florida, he said is getting nutrients into the sandy soil. "It has no fertility at all," he said. "And growing full-size slicing tomatoes is particularly hard because they have to stay in the ground for such a long time that they can use up all the nutrients we've worked in before they ripen." Beddard spreads spent compost from mushroom farms over his land before crops go in. When they are harvested, he plants his fields in cover crops of sorghum and cowpeas, which add nitrogen and also help ward off the nematodes that conventional growers kill with toxic fumigants like methyl bromide and methyl iodide. He pointed to a field that was ready to be prepared for planting. It looked like a stubbly hayfield that had been cut but not baled. "The cover crops were hip high before we mowed it," he said. "There's a huge amount of organic matter there." Conventional farmers allow their fields to grow up in weeds during the off season, which they then kill with herbicides. Beddard simply disks his cover crops into the soil. And where a conventional farmer would grow tomatoes in the same field year after year, Beddard practices crop rotation—tomatoes, peppers, eggplants, salad greens. He says that his yields are lower than his chemically dependent colleagues, sometimes significantly, but he more than recoups the differences in yields through the higher prices he can command for organic produce. "I go down there to Immokalee and I envy those guys with their plants just hanging with tomatoes," he said. "But I probably make more than they do per acre."

Beddard, who is fifty-five years old, grew up a city boy in Pittsburgh. His parents were shocked when at age sixteen he announced that he wanted to become a farmer, but they allowed him to study horticulture at Delaware Valley College in southwestern Pennsylvania. In those days, the school not only did not teach organic farming techniques, it actively frowned on them. But even as an inexperienced student, Beddard thought that there was something viscerally wrong with using poisonous chemicals to grow food we would eventually put in our mouths. "I was viewed as a renegade with a hippie philosophy," he said.

Upon graduation, Beddard discovered that there wasn't any job he could get with his horticulture degree, aside from going into agribusiness or selling agricultural chemicals, so in 1988 he and his wife, Chris (who died in a car accident in 2004), bought twenty acres in Pennsylvania, five of which were tillable. Over the next ten years, Lady Moon grew steadily, selling first to local health food stores and eventually to larger supermarket accounts. The Beddards purchased more land as business increased and expanded from a two-person organization that dragooned the Beddards three children during peak periods to one that now employs 150 workers year-round. "I was fortunate that I hit the organic scene just as it was starting to take off," said Beddard.

By the late 1990s, the Beddards had what seemed like an ideal agrarian situation. Although they worked long hours throughout the spring, summer, and fall, they were making a decent living and they had the entire winter off. When Beddard suggested that they buy some land in Florida to offset the risk from having all their crops on one farm, his wife wisecracked, "Yeah, and then we can work 365 days a year."

Despite the dire predictions of experienced Florida growers, Beddard had a bountiful harvest his first season in the South. He felt smug until he read in *Packer* magazine, a trade publication for the vegetable industry, that growing conditions in Florida that winter were the best they had been in a century. Then a few years later, a single storm, Hurricane Jeanne, damaged his Florida fields, and then moved up the coast, hitting his farms in Georgia, before drowning out what remained in his Pennsylvania fields.

In 2008 his buyer from Whole Foods Market came to Beddard and said that the natural grocery chain intended to sign the Coalition of Immokalee Workers fair food agreement. They wanted to know if he would be willing to pay harvesters an extra penny per pound and comply with the coalition's other terms. That request presented him with a problem. As a matter of policy, Beddard pays hourly wages—there is no antiquated per-pound piece work at Lady Moon. In addition, he provides free housing for his workers when they move north to Georgia and Pennsylvania for the summer. (They are responsible for their own accommodations in Florida, where they live for most of the year with their families.) He consulted his accountants and discovered that he was already in compliance by a comfortable margin with the demands of the Campaign for Fair Food. He laughed, shaking his head. "I mean, come on, we're talking about a *penny* per pound. What's a penny a pound to these big producers? What's it to me? Nothing. It made no sense to me why they fought so hard and in doing so gave the coalition all the ammo they could have ever asked for. I told one of them, 'Give them the damn penny per pound and they'll be off your back.'"

We arrived back at his packinghouse, where Beddard carried on a conversation about laying some irrigation lines for a new field with his field foreman, a slight, mustachioed Latino whose features were hidden in the deep shade provided by the brim of his straw cowboy hat. As I turned to leave, he told me, "Organic farming in Florida can be a bitch," he said. "But it can be done."

THE TEACHER

Barbara Mainster cracked open a door in a building behind the Redlands Christian Migrant Association's head office in Immokalee, turned to me, and put her index finger to her lips. "Shhhhhhhh!" She led me into a darkened room. It was silent and the cool air inside provided welcome relief from the numbing humidity of an overcast autumn day. Even in the dim light, the room and all its furnishings abounded in reds, yellows, and blues. It was spotlessly clean. In a whisper, Mainster introduced me to two Hispanic women who both looked like they were in their late twenties or early thirties, Hilda Enriques, in a rocking chair, and Francesca Sota, who was stretched out on the floor. Both women cradled infants in their arms. One other baby slept in a crib. "We usually have eight babies in this room," said Mainster, a gray-haired seventy-year-old grandmother. "The others have gone home for the day." Enriques had been a caregiver at Redlands for five years; Soto for seven. Before that, both women had labored in the same fields where the mothers of the children in their arms were working at that very moment. "They have walked in the same shoes as the babies' parents," said Mainster.

And that is the key to the success of Mainster's association, which provides free or low-cost child care and early education to the children of migrant farmworkers and other rural, low-income families. The organization began in 1965, when a group of Mennonite Church volunteers decided to provide daycare for the children of workers who lived in two migrant camps in Redlands, an agricultural district about twenty miles southwest of Miami. The goal was to keep the kids safe and out of the fields.

Initially, it seemed that the Mennonites' plan was fated to be just another well-intentioned charitable effort that fizzled. The founders opened the centers, but no one came. The immigrant mothers were not comfortable leaving their tiny children with white Americans who spoke no Spanish or Haitian Creole and had little understanding of the parents' cultures. Only when the association began hiring from the migrant community itself did the centers begin to fill.

Daycare for the children of immigrants provides a double-edged benefit. Kids who might otherwise be hauled out into the fields or warehoused by the dozen in filthy trailers supervised by the uneducated wife of a crew boss are given a clean, safe environment and acquire the basic skills necessary to enter the American school system. The women who care for them are able to leave the fields to work in comfortable, secure surroundings. They are encouraged to continue their education and earn living wages, creating pockets of upward mobility in the migrant communities.

When Mainster, who has been executive director since 1988, joined the association in 1972, it had three centers in one county, with an enrollment of seventy-five kids. Today, thanks to her single-minded drive, which has not diminished an iota over nearly four decades during which she raised four of her own children (three of them adopted), the association has more than eighty centers and charter schools in twenty-one counties. It serves eight thousand children, making it one of the largest nonprofit child-care programs in the United States, and employs fifteen hundred caregivers, most of them Latinas. Redlands kids get a hot breakfast, a hot lunch, and an afternoon snack and learn enough English to enter the American school system.

Mainster and I strolled into a shaded courtyard that houses three separate age-appropriate playgrounds with plastic tunnels, slides, tricycles, swings, a playhouse, and a pretend gas station and café, all painted in riotous primary colors. The space, located in the former Sunday school of a Baptist church, now provides care for 180 children. An additional 220 first- to sixth-grade students attend a charter school run by the association on an adjacent piece of property. "We want to keep them with us as long as possible," said Mainster.

Mainster credits much of the success of Redlands to her philosophy that there are good people and bad people in every profession, including Florida agriculture. Mainster works closely with major growers—Michael Stuart, chief executive officer of the Florida Fruit and Vegetable Association, a trade group, was president of the Redlands board of directors. "Agriculture is a very important, well-connected force in this state, and they can lobby very effectively for funding and other things we need," she said.

A sign in Mainster's office reads "We don't believe in miracles. We rely on them." In truth government grants through programs such as Head Start and federal child care grants cover 85 percent of the association's $56-million annual budget. Donations and charter school fees make up most of the rest, with only 2 percent coming from parent fees. Despite being called the Redlands Christian Migrant Association, the group is completely secular. "I don't know why we stick with the name," Mainster said. "We are no longer based in Redlands. We are not affiliated with any religion. We work with nonmigrants as well as migrants. And we are a nonprofit organization, not an association."

Ultimately, Redlands' goal, according to Mainster, is to level the playing field for the children under their care. A child of English-speaking native-born American parents has a vocabulary of about three thousand words at age three, she explained. The child of an uneducated, non–English speaking mother has only five hundred. Some Redlands students in the sixth grade, she said, have never been out of Immokalee, except on school field trips. "They start out at a huge disadvantage."

A dozen years ago, a Guatemalan boy who came to one of their centers at about age two certainly faced more than his share of challenges. His mother was illiterate and spoke only her native Amerindian language and some fragmentary Spanish. The child was completely nonverbal. Mainster arranged for him to be tested for hearing loss and mental disabilities. He had none. Caregivers continued to work with him, and slowly he began to speak. By third grade, he was not only fluent in Spanish and English but was reading at levels deemed age-appropriate by the state. In sixth grade, he asked to speak to a guidance counselor and came into her office carrying a thick envelope from the Florida Board of Education written in English and aimed at educated American parents. Consulting a document, he said, "It says here that it's time that I do a little career exploration."

"This kid would have been considered gifted in an upper-middle-class school setting," Mainster said. The boy is in high school now, and Mainster intends to make sure that Redlands gets him scholarship money and anything else he needs to attend college and, who knows, maybe get a fair shot at the piece of the American dream denied to his parents.

THE BUILDER

Hurricane Andrew made landfall near Homestead, Florida, at about five o'clock in the morning on August 24, 1992. The powerful storm's 150-mile-per hour winds ripped into the coastal area just south of Miami, slamming into a dilapidated, county-owned trailer park that for two decades had served as a "temporary" labor camp for the migrants who picked citrus, tomatoes, and other crops grown in Dade County's gravely soil. Miraculously, no one who lived in the park was injured. At that time of year, most migrant workers are in northern states. However, all but two of the four hundred trailers were demolished, reduced to heaps of splintered two-by-fours, twisted aluminum siding, and ripped-apart furniture. Andrew left the 154 families who lived in the camp at the time homeless. That hurricane turned out to be one of the best things ever to happen to Florida's farmworkers.

In the aftermath of Andrew, South Florida found itself awash in offers of federal disaster relief funding. Farmers desperately needed shelter for the workers that would soon be arriving to pick the winter's harvest. The board of the Everglades Community Association, which managed worker housing in the county, hired Steven Kirk to oversee reconstruction. It was a classic case of the right person for the right job. Kirk, who had studied public policy at Duke University, became a passionate advocate for farmworker justice after spending a summer in the mid 1970s interviewing vegetable pickers in North Carolina under the supervision of the oral historian and author Robert Coles. One day he stumbled across a couple of African American men running down an unpaved road. They told him that their boss had tied them to a tree to prevent their escaping. Kirk rented them rooms in a motel until authorities came.

Kirk spent the early part of his career knocking around Washington, DC, working for various farm laborers' groups and other antipoverty organizations, gaining insight into how to manipulate levers of power and loosen purse strings in the nation's capital. Upon arriving in Florida, he saw two courses of action for the Everglades Community Association. They could simply replicate the ugly, crime-ridden old camp by acquiring a few hundred replacement trailers and slapping them down in straight barrackslike rows on cement pads, or they could do something no one else had attempted: build a functioning farmworkers' community.

Today, Everglades Farmworker Village, as the one hundred and twenty acre development that sprang up on the ground occupied by that old trailer park is called, is one of the country's largest farmworker housing projects. In one of the electric golf carts that provide the primary mode of transportation for village maintenance people and other employees, Kirk gave me a tour. Short, wearing jeans, with mussed, thinning hair, Kirk is in his mid-fifties. As we purred through a pleasant network of curving streets bordered by palm trees, he told me that the community is home to nearly two thousand mostly Hispanic workers whose average family income is between $16,000 and $18,000. The 493 housing units, pastel stucco over cement block, are either standalone single-family structures, side-by-side duplexes, or two-story townhouses. A couple of dormitory-style buildings provide accommodations for 144 single men. The streets have curbs and gutters, and the landscaping is immaculate. The community has its own ten-acre park and soccer fields. A small grocery store, a branch of a larger Hispanic supermarket in Homestead, provides a wide range of traditional products at reasonable prices. A Community Development Credit Union maintains an office here, and workers can get fairly priced loans, open bank accounts, and make other financial transactions so they are not gouged by check-cashing companies and costly wire transfers.

Kirk made a deal with Mainster's Redlands organization to run three day care centers at Everglades serving three hundred preschoolers. An additional 250 older kids participate in organized after-school activities. The community has a space for religious services, a community hall for wedding receptions and *quinceañeras* (celebrations of a Latina girl's fifteenth birthday), a computer lab, two self-service laundries, and a health clinic. "There are a lot of people involved with low-income housing whose attitude is, they are just going to tear the places apart, why make them nice," Kirk said. He takes the opposite view, believing that if you have high expectations of tenants and give them quality accommodations, they will respect them. "And that has proved to be true," he said.

In the world of migrant housing, the lowest of the low are single men. They are typically the ones relegated to sleeping ten or twelve to a trailer in places like Immokalee—if they have any place to stay at all. The backcountry of Florida is pocked with makeshift encampments of single workers who cannot find shelter. Even farmworkers with families shun their single brethren, associating them with loud music, drunken rowdiness, and unwanted interest in teenage daughters. But Kirk was determined to make a place for single men in Everglades Farmworker Village.

He stopped the golf cart and opened a gate that led into a shady courtyard surrounded by a U-shaped building whose facade was regularly interrupted by doors, giving the effect of a motel that had turned inward on itself. Gazebos and clusters of benches, chairs, and tables filled the courtyard. The principle, Kirk said, was to provide the men with a space to mingle and socialize outdoors that would also contain their activities and provide a measure of control over who entered the compound. Kirk opened the door to a unit, exposing us to a puff of air-conditioned air. The living room consisted of a heavy wooden table with benches set on a spotless linoleum floor. Off that room was a kitchen with a stove and two refrigerators. "Eight guys share this space, we want them to have room to store their food," Kirk explained. Four bedrooms extended off the main living area, each with two built-in twin beds. Toilets and showers were in separate rooms.

The quarters reminded me of the on-campus apartment where my college-age daughter lived with four friends, only the workers' was more spacious and cleaner. At $175 per person a month including all utilities, the bachelor accommodations were a steal compared to the trailers I'd seen in Immokalee. And no one complained about the single guys. "We get more complaints about teenaged kids of married couples," Kirk said.

When designing the village, Kirk sat down with prospective residents and asked them what features they wanted to see in their dwellings. Women wanted to have hook-ups for washers and dryers, plenty of storage space, and large kitchen windows so they could keep an eye on their kids playing in the yard while they prepared meals. Men wanted parking places installed tight up in front of the houses to deter anyone who might want to vandalize or steal their vehicles—in many cases the only asset the family possessed. Everyone wanted to save money on electricity bills, so houses included fans in every room and specially designed windows to ventilate homes, limiting the need to run expensive air conditioners. A gated entrance and night-time security were also on the workers' wish list.

In return, Kirk and the board, which included residents, had a few demands of their own. "We practice tough love," he said. Some might say it's paternalistic or downright authoritarian. But it works, Kirk insists. Quiet must prevail after 11:00 in the evening. Single men can have no overnight guests. Vehicles must be parked in designated places. No pets are allowed. No do-it-yourself paint jobs or landscaping projects are permitted. There are no clotheslines, a rule Kirk explained by saying that aesthetics are as important as any other issue to maintaining a sense of pride in the community. The final rule is you have to pay your rent, which is capped at one-third of a family's income. Government subsidies make up the rest, if necessary. "We don't evict people who are unable to pay. We evict them for refusal to pay," said Kirk. "If rent is affordable to people, it becomes a priority for them."

Early on, Kirk faced some competition for funding from farmers who wanted to build housing for their workers on their own land. Recalling incidents of being run off property by county sheriffs back in his student days when he was trying to interview the children of North Carolina farmworkers, Kirk adamantly opposed employer-built housing. "I don't want them in control," he said. "If you live down some dirt road where there's an armed security guard to keep people out, problems can develop. Here, if someone from Florida Legal Services wants to come down and meet with you, there's an office set aside. The opportunity for involuntary servitude in this community is pretty slim. There is always someone to reach out to."

In the 1990s, Kirk, a self-professed workaholic, spent many nights at his desk until ten o'clock at night, a muted TV in one corner of his office tuned to CNN providing a link to the outside world. Equal parts zealous missionary for the oppressed and hard-driving real estate tycoon, Kirk had no family, few outside interests, and a very neglected girlfriend. His $50-million project was up and running. The creative, challenging part had been successfully completed, and suddenly Kirk found himself growing antsy. He went to his board of directors, which consisted entirely of local people whose interests were focused on county politics, and spelled the situation out. It was fine with him if they just wanted to run Everglades Farmworker Village. "I told them, if that's the case, you need a property manager." That job was not intellectually challenging enough for him. On the other hand, he said, the group could take what they had learned in building the village and try to spread affordable housing to farmworkers statewide. "The situation is very, very bad upstate," he explained, saying that if they wanted to expand, he was their man. They told him that they were prepared to take the next steps.

Over the ensuing decade, an umbrella organization called Rural Neighborhoods, with Kirk as its president, built developments in Immokalee, LaBelle, and Okeechobee in the southern part of the state, in Winter Haven in the center, in Ruskin in the west, and Fort

Pierce in the east. With low-interest loans from the U.S. Department of Agriculture and through the federal Low-Income Housing Tax Credit that encourages private investors to make equity investments in affordable housing projects, Rural Neighborhoods spent more than $200 million to build thirteen hundred units that house more than forty-five hundred tenants, mostly farmworkers. It was an astonishing accomplishment. Kirk credits his knowledge of how governmental agencies work and his acumen about what politicians and officials need for his ability to get his projects funded. A conservative governor taking heat for working conditions in Florida fields might be eager to allot housing funding to migrant housing as a goodwill gesture. "We let them co-opt us; and we co-opt them," said Kirk.

Other affordable housing advocates have criticized Kirk for the high quality of his housing. "Why spend $100,000 per unit for architect-designed cement block and stucco structures when you can get a manufactured home for $60,000?" they complain. Kirk is utterly unrepentant. Rural Neighborhoods has built the best apartments available, period, in places like Immokalee, LaBelle, and Okeechobee, he said. "We've changed the perception of farmworker housing. I would be happy to live in any of our developments."

There is also an element of canniness in building attractive communities that look nothing like stereotypical farmworker housing. Given their propensity for being blasted into kindling with each passing hurricane, manufactured homes (glorified trailers, in the view of many) have an unsavory reputation in Florida. "Yeah, maybe we could house more people for less money in the short run," said Kirk. "But the political reality is that quality housing sells better to local lawmakers." It is also a financial as well as a political reality. If a lender ever had to foreclose on a Rural Neighborhoods' property, he could sell it to a landlord who could fill it with eager renters, not marginalized farmworkers, which makes it more likely that a bank will extend credit to Kirk.

Unlike most heads of charitable organizations, Kirk lives with the possibility of bankruptcy. Rural Neighborhoods may borrow at low interest rates through government programs, but the money it invests has to be paid back like any mortgage. Rents in its developments are set at a level that allows Rural Neighborhoods to break even, but higher than expected vacancy rates can quickly turn break-even budgeting into a losing proposition. "We are a risk-taking organization," Kirk said. "We are doing multimillion-dollar deals. We guarantee loans. We could fail. But my view is that Bob Dylan thing, 'He not busy being born is busy dying.'"

Just prior to my visiting him in 2010, Kirk did something completely out of character: He took off three days in a row. On the previous weekend, the fifty-five-year-old lifelong bachelor had gotten married for the first time. But having entered into that state hadn't done much to lessen his pace. He had 281 new units—$35 million worth—under various stages of construction and scheduled for occupancy within a year, all of which had to be paid for. The recession was making it tough even for providers of homes for the lowest of the low. Plus, Kirk was concerned that the immigration crackdown and resulting fear among migrants might suddenly leave him with unexpected vacancy rates. But none of those worries were slowing him down. "There's need out there," he said. "And if no one else is going to fill it, I have to step into the void." He shrugged philosophically and added, "And even if we fail, those new units will still be there."

If Kirk is wistful about anything after three decades as a farmworkers' advocate, it's that there's still a need for people like him. "We set out trying to change agriculture," he said. "People like Barbara Mainster and I have changed conditions for some workers, but we haven't changed agriculture."

TOMATOMAN

It was a little after four o'clock in the morning when a tractor trailer careened into our lane. Tim Stark and I were in his pickup truck on I-78, about an hour and a half west of our destination, Greenmarket Union Square in New York City. In the back, we had five hundred pounds of vine-ripened tomatoes. Somewhere ahead of us in the darkness, Stark's other, larger truck, a box truck, lumbered along carrying an additional thirty-five hundred pounds. With any luck, most of those tomatoes would be sold that day to customers at Greenmarket as well as to a couple of dozen of the top restaurants in Manhattan and some specialty produce distributors. But first, we had to survive the journey into the city from Stark's farm in rural eastern Pennsylvania.

"You always have to watch out for the swerving trucks at this time of the morning," Stark observed casually, as the one that had nearly cut us off veered back into its own lane and then swayed onto the shoulder. Seizing the moment, he gunned his engine, saying, "You want to pass them as fast as you can. And pray."

I was riding shotgun beside Stark that morning in an attempt to understand how he has succeeded where hundreds of like-minded idealists have failed. In the mid-1990s, the Princeton University graduate abandoned his city-bound lifestyle and the consultant's job that provided an income while he tried to break into the business of being a professional writer. He dreamed of earning a livelihood by growing and selling tomatoes from a rocky few acres of marginal land on a hillside behind his mother's house. Although he had gardened there in the past, he had zero experience as a farmer. "I just wanted to make a living where I grew up," he said.

Stark, who has been called "the Titan of Tomatoes" by *New York* magazine, though most people know him as simply "the Tomatoman," sells his famously delicious tomatoes in the New York City area, avoids agricultural chemicals, and provides his staff with free housing, livable wages, and benefits. Eckerton Hill Farm, as his company is called, shows that with clever marketing and a hell of a lot of work, there is still a place for small farms that grow great-tasting tomatoes for regional markets in a manner that is sustainable both for the environment and for the people who work the land. Given the vulnerability of Florida's industrial tomato agribusinesses to competition from Mexico and the wild price swings in the commodity market, it could be argued that Eckerton provides a more viable business model for commercial tomato growers, a model that is sustainable in more ways than one.

Success, however, did not come easily. "I took a long strange route to get here," he said. "For the first ten years, I had my head down in the mud trying to make this thing work. It was hard on my marriage and hard on my family." Instead of a bucolic return to his rural roots, Stark found himself embroiled in a decade of nonstop crisis management, a story he relates in his 2008 memoir, *Heirloom: Notes from an Accidental Tomato Farmer*.

Eckerton was an accidental farm. It came into being when Stark's landlord summarily evicted three thousand tomato seedlings—sixty different varieties—that Stark, seduced by the inexpensive nonhybrid heirloom offerings from other gardeners in the annual catalog of the Seed Savers Exchange, had too ambitiously grown on the roof of a Brooklyn brownstone in anticipation of moving them gradually to his mother's property. He was forced to load his decrepit Toyota pickup truck and convince four of his professional urban friends that transplanting tomatoes with trowels into two acres of rocky terrain for no pay would be a great way to spend a pleasant spring weekend.

That first summer was one of the few times the elements conspired to help Stark's nascent farm. Rains came in gentle, weekly intervals, delivering just the right amount of moisture. "I was really, really lucky," Stark told an NPR interviewer. "If we'd had a dry year that first year, because I had no irrigation, I wouldn't be here. I would be back in Brooklyn thinking up my next scheme." Stark, then thirty-three years old, spent the early part of the season frenetically pulling the weeds—quack grass, purslane, wild mustard, Canadian thistle, wild carrot, and crabgrass—that also benefited from the perfect growing conditions, while he was falling further and further behind.

Despite competition from the weeds, Stark's crop thrived, ripening before other area farmers' did. The timing couldn't have been better. It was that period of the year when locals were salivating for the first real summer tomato. Stark was certain they would be willing to pay premium prices. With visions of the many uses to which he would put his profits (the first being basic survival), Stark began to harvest his misshapen, multicolored cornucopia—heirlooms whose names say it all: Black Krim, Black Seaman, Casady's Folly, Vintage Wine, Speckled Roman, Pink Lady, Purple Calabash, Extra Eros Zlatolaska, Aunt Ruby's German Green, Cherokee Purple, Green Zebras, Yellow Brandywine, White Wonder, Sungold, Radiator Charlie's Mortgage Lifter. Unfortunately, buyers at the local produce auction had a clear preference for smooth, round, red beefsteak tomatoes, a position similar to the stance of the Florida Tomato Committee. They wouldn't so much as bid on Stark's motley harvest until the auctioneer dropped the price to less than what it cost Stark to grow and pack them.

The only hope for Eckerton lay in New York's Union Square farmers' market, one of more than fifty markets operated throughout the city as part of the Greenmarket program. Years of living in the city had convinced Stark that the food-savvy citizens of Gotham would be eager to embrace his off-beat harvest. He tossed a meager fifteen pounds of tomatoes into his pickup and made his first two-and-a-half-hour predawn run into town on I-78. He set up his table, covered it in tomato-themed fabric that his then girlfriend and now wife, Jill, had bought, and crossed his fingers. A woman wearing the tight-eyed scowl of a skeptical New Yorker approached and began poking and squeezing his blemished offerings. She brusquely informed him that a young man tending another stall in the market had tried to convince her that his tomatoes were field grown. "I took one look at them and said, 'Yeah, I just fell off the turnip truck myself,'" Stark wrote in his memoir. The woman bought some of Stark's tomatoes and asked if he would have any the following week. He assured her he would, but seven rain-soaked days later, he could harvest only twenty pounds to bring into the city, barely enough to fill a single peck basket. Customers descended on the basket like "a single many-tentacled organism," Stark wrote.

When the rain finally stopped, Stark's tomatoes began to ripen faster than he could pick them. The night before market, he'd load his truck until one o'clock in the morning, filling the bed and then the cab with flats, leaving just enough space to slide in behind the wheel. When he arrived at market, customers bought at such a pace that he often didn't have time to set up his tables with Jill's pretty tablecloths. Shoppers couldn't get enough of his heirlooms, the uglier, more cracked, and more beaten up, the better. He sold out every time he went to town, and Eckerton limped along—barely. The venerable pickup truck expired under a load of tomatoes, and Stark had to rent a U-Haul, stowing excess tomatoes in his mother's Subaru station wagon, which Jill drove into town.

Stark still had years of twenty-hour workdays ahead of him, picking and weeding when the sun shone, packing by night, and heading into New York before dawn after a few stolen hours of sleep. It was a good thing that he possessed inexhaustible stores of stamina, developed in part by having undertaken epic bicycle trips to the West Coast, to a friend's wedding in Detroit, and across the Alps. On the farm in those early years, he dealt with drought, frosts, flooding rains, legal actions by angry neighbors, voracious tomato worms, grasshoppers, deer, groundhogs, supercharged weeds, malfunctioning machines, and a revolving cast of workers who would stay just long enough to realize that farm labor was hard, hot work and then disappear at the very moment when Stark needed all the help he could get. Often at the end of a season, he promised himself that it would be the last. But a few months later, with the first lengthening days of spring, Stark inevitably found himself in his greenhouse putting tomato seeds into potting mix. "I became neurotic," he said. "You would, too, if you had to make a living this way. Dry spells, bugs, freezes, even a single rainy Saturday when no one comes to market cost me money I didn't have." Stark mentally punished himself for months after trapping and killing a woodchuck that was chomping down his seedlings. He took to setting his alarm clock for midnight and four o'clock in the morning so he could check that mice and voles hadn't gnawed tiny holes in his irrigation hoses. When sleep came, it was interrupted by vivid dreams of snow falling in thick flakes on his unprotected crops. He'd wake with a start, only to realize that it was a warm summer night. But when he tried to go back to sleep, he would start fretting about losing his crops to late blight—a very real threat that he could do little about.

A big breakthrough came when he hired a crew of experienced Mexican immigrant workers, veterans of agribusiness operations in California and Florida who had moved to Pennsylvania and settled into the Hispanic community in nearby Reading. Although his new employees found his methods crude, amateurish, and borderline amusing (watering with hand-carried milk jugs? manually pulling weeds? not spraying pesticides?), they were dependable and brought a necessary stability to the Eckerton workforce. Not only could they pace themselves to put in long days in the humid fields, they repaired machines that broke down without coming to Stark and were skilled handymen, able to perform all the chores a small farm demands. Days became less frenetic. Rented U-Hauls became things of the past when Stark purchased a genuine refrigerated produce truck (albeit one with 191,000 miles on it and a refrigeration unit that blew warm air) and painted tomatoes on it above the farm's slogan: "Home of the Tomato People." He was able to rent an additional patch of land with a house that he and Jill moved into. Then, two years ago, he finally became a landowner after purchasing a fifty-eight-acre farm. Today, Eckerton employs fifteen workers with a payroll that can top $12,000 a week during peak times of the season. It is home to thirty thousand tomato plants that Stark starts in his own new greenhouse. He and Jill and their two middle-school-age daughters live in a neat cape in Kutztown, a college town about fifteen minutes from his farm. "As long as I have tomatoes, there's money to be made," he said. "But I still get nervous when tomatoes aren't coming in. Early in the season when we're just planting and staking, money is pouring out of this place."

Stark can become wistful when remembering the early years. As the public face of the farm, he spends more of his time managing, marketing, and selling than he does planting, picking, and packing. His worries are not so much about weeds and water as they are about making payroll and dealing with a mid-six-figure bank loan. Nonetheless, when I met him the day before that predawn haul into the city, I immediately understood how idealists who are not familiar with the vicissitudes of agriculture can succumb to romantic illusions. Sunlight shone on the rolling fields with a crisp clarity that only comes on cool autumnal days. Stark, who still has the compact muscled build of his college wrestling days, was in black jeans, a black sweater, and dusty work boots. His light brown hair, which was uncannily free of gray given his age and the stress he has endured, was a tangle of waves that badly needed a trim, and several days had obviously come and gone since a razor had passed over his face. As Stark walked out to a field to check on the progress of a crew of about a dozen workers picking cherry tomatoes and loading them onto flats beneath the shade of a blue plastic tarpaulin, a fox cavorted on the border of the forest, pouncing occasionally on an unfortunate field mouse or meadow vole. "I still think of myself as more of a gardener than a farmer," he said. "My secret is that I do everything wrong."

By the standards of the fresh tomato agribusiness in Florida, he does. Eckerton's crops are grown for their taste, not for a three-week shelf life and the ability to withstand the rigors of factory packing and shipping. His tomatoes are harvested at the moment of full ripeness, not when they are hard and green. What gets picked on one day is sold the next. Instead of growing in perfect parallel plastic-covered rows of sterilized, weed-free soil, Stark's crops make do with straw mulch and engage in Darwinian competition with weeds. "We put our plants through hell," he explained. "Maybe that's what gives them taste and character." Where factory farms focus on single varieties of one crop, Stark grows more than one hundred varieties of tomatoes alone, both heirloom and hybrid, in addition to peppers, peas, broccoli, carrots, sweet potatoes, leeks, onions, potatoes, herbs, lettuce, raspberries, kale . . . and the list goes on.

He isn't organically certified because, he admits, he is not organized enough to keep up with the paperwork required by the U.S. Department of Agriculture. In fact, he's always rummaging around in search of some lost possession—his cell phone, his wallet, his pen, his to-do list, his checkbook. His computer is regularly on the fritz for one reason or another, making it difficult for customers to contact him by e-mail. "If they want to reach me, they'll call and leave a phone message," he said. He keeps none of the detailed seasonal planting charts or harvest records that most commercial growers—even those much smaller than Stark—maintain. Everything is in his head. "His mind is like a calculator," one of his workers told me. "He remembers everything."

Besides, he reserves the right to apply inorganic chemicals if he must to ensure the survival of his business. In fourteen years, that has happened precisely once, in 2009, when the northeastern states were hit with an epidemic of late blight that would have wiped out his all-important tomato crop. When it became time to spray the vines,

Stark insisted on doing it himself rather than expose his workers to potentially dangerous fungicides. He uses no chemical fertilizers, instead relying on ground-in cover crops, spent compost from nearby mushroom growers, and manure from the horses of a neighbor who is more than happy to have someone take it off his hands. Stark demonstrated the extent of his pest management program to me by stopping mid-sentence and nonchalantly plucking a three-inch tomato horn worm off a plant as we walked by and grinding it under his heel. "I still have a gardener's mentality," he said.

The money he saves on chemical fertilizers and pesticides is more than counterbalanced by the costs of the extra labor needed to control weeds and pick crops by hand. And not only is his workforce larger than it would be if he resorted to the tactics of industrial agriculture, it is more fairly compensated. Stark says that paying by the amount picked is simply illogical. Different varieties of tomatoes yield different quantities of fruit. If you're working a row that's closer to the packing station than another worker, you will be able to bring in more tomatoes. And workdays at the farm involve numerous duties. Mornings may be taken up with picking, afternoons with packing flats. Some days there is no harvesting at all, and attention shifts to weeding, tilling, or maintaining farm machinery. During the 2010 season, wages at Eckerton started at $10 per hour for college-age interns (above Pennsylvania's minimum wage of $9.35) and topped out at $15 an hour for experienced employees. Stark pays bonuses at the end of the year. He provides free housing for his help (the interns are expected to "pay" for their board by milking Eckerton's single nanny goat) in one of two houses on Stark's properties. His employees are provided with workers' compensation and unemployment insurance. Stark has tried to find a health insurance company that will cover them, but so far none has been willing to so much as consider it. "I'm proud to pay a living wage," he said.

Doing so has advantages. Stark's second-in-command, Wayne Miller, has been with him since 1998, and he and his wife recently purchased a house on five acres of land three miles down the road from the farm. Other members of his crew, which includes both men and women and is comprised equally of Mexican immigrants and native-born Americans, arrived in 1999 and 2000. "Farming well is not dummies' work," he said. "It would be nice if we all respected it more."

Stark nonchalantly invited me to accompany him over to a house where some of his employees lived. He had pressed some apples a few days earlier and wanted to pick up one of the jugs of cider he had stored in an extra refrigerator there. We bounced off the paved road onto a long gravel driveway bordered by Eckerton's trademark ranks of shaggy tomato plants and weeds. The lane ended at a restored nineteenth-century log farmhouse on a hillside overlooking a pond. There were three bedrooms upstairs, each serving as a private space for either a single person or a couple. The downstairs was dominated by a bright, open kitchen with a wall of windows and a snug living room. The place was overdue for a renovation and could have used some general sprucing up but was in no worse shape than other farmhouses in the area that rent for $1,000 or more per month.

Despite its benefits, working for Stark can be trying. In describing himself, he often uses the adjective "cranky" and admits to throwing the occasional temper tantrum, especially during late-night sessions in his packinghouse on the day before market. "I'm not the nicest guy. I find myself saying to staff members, 'Hey! You can pack and talk at the same time.'" He once caught one of his workers texting friends when they had a line of customers at Greenmarket. She was thereafter demoted to field and packinghouse duties. Looking over full boxes ready to be loaded onto the truck, he became visibly angry when he didn't see enough cherry tomatoes to fill advance orders from chefs. "They must have more, somewhere," he said, moving a few stacked boxes to one side. "They better have or there'll be trouble." At the market, he pulled Miller aside and pointed to a quart box of salad tomatoes topped by a fruit bearing a black, oozing gash and icily said, "This does not make me happy. We have to watch how they pack these boxes." And even though a month had passed, Stark was still grumbling about a review in *New York* magazine that rated one of his Cherokee Purple tomatoes eighth out of eighteen heirloom varieties its reviewers taste tested. "They said they went to each grower and asked for his best. Well, they certainly didn't come to me. I would have never given them a Cherokee Purple. So who did they go to? I don't know. At least it didn't do any harm to my business. My customers still like my tomatoes."

Miller, who comanages the farm, concentrating on supplying the markets (Ignacio "Nacho" Baltazar oversees field operations), credits Stark's success to "a decent dose of obsessiveness." Miller sees his role as part buffer between Stark and the employees and part translator of Stark's instructions, which tend to be delivered in staccato half-sentences that have no immediately obvious linkage to each other. Miller explained, "Tim will come in and go on for five minutes, and when he's gone a worker will turn to me and say, 'What just happened? Did we just decide something?' and I will have to spell out what Tim wants done. Plus, he is a very emotional person and can let problems build up. He'll stew for a week or two, and then it will come out in a burst. But there is an endearing charm to Tim. He's engaging, like an absent-minded professor or an obsessed artist."

Stark's longtime customer, Dave Pasternack, chef owner of ESCA, an acclaimed seafood restaurant in Manhattan, summed up his personality more succinctly, saying that he was "one-fourth farmer, one-fourth storyteller, and half mad."

The key to success for farms such as Eckerton, Stark says, is clever niche marketing. His mantra is "eliminate the middleman." In so doing, he contends that small farmers can become price makers, not price takers like those who produce commodity crops and sell into vast distribution networks. Even the largest Florida tomato grower is at the mercy of violent swings in market prices that are completely beyond his control—a freeze sends prices soaring or a stretch of good weather can result in too many tomatoes ripening at once and flooding the market. And if doing away with middlemen means hauling yourself out of bed at three o'clock in the morning three or four days a week from May through November, so be it.

The sun had yet to rise when Stark and I emerged from the Lincoln Tunnel into Manhattan's West Side. In the hours just before dawn, New York's sidewalks were eerily devoid of pedestrians. Instead of horn-blaring yellow cabs, side streets were crammed with double-parked box trucks and delivery vans, each tended by a driver in the process of off-loading pallets and cardboard containers. Stark and I pulled up in front of Becco, a theater-district restaurant owned by the renowned Lidia Bastianich and her son Joseph. A half-asleep kitchen underling unlocked the door, and we carried in two hundred pounds of assorted salad tomatoes. A few blocks south at ESCA, Stark and I lugged eight cartons of tomatoes down into the subterranean kitchen. Those errands out of the way, we headed downtown to Union Square, pulling up just as the eastern horizon began to brighten.

Miller and another Eckerton employee had driven into the city in the farm's big truck, a faithful 1991 Hino that has logged 250,000 miles ferrying produce from Pennsylvania to Manhattan. They had already set up the booth, covering the tables with the farm's trademark colorful tablecloths, which were immediately obscured by boxes overflowing with vibrant tomatoes and chiles—red, yellow, orange, purple, green. Even though it was still well before opening time, a cluster of women speaking in lilting Caribbean accents jostled for prime positions in front of one of the tables. As soon as an Eckerton worker put out a carton filled with bright yellow Grenada seasoning peppers, which are flavorful but not particularly hot, I understood why: Within minutes, every pepper had vanished, and then disappeared again when the carton was refilled. The day's first customers walked away toward the subway with bulging shopping bags. Stark had yet to sell a tomato, but the birth of this small sideline to his main business demonstrates his niche marketing philosophy in action. Several years ago, a Grenadian woman approached him at the market and held out a single pepper and told him that he would have eager customers if he grew them. A less savvy farmer might have smiled and tossed the pepper on the ground as soon as she turned away. Stark saved the seeds and by the following July had added another popular item to his product line.

As the morning wore on, the market filled with shoppers. Two interns manned the front of the Eckerton stand, handing over produce, making change, and restocking tables as customers picked up cardboard containers that sold for $3.25 to $4 depending on the variety and size, a dollar more expensive than the cheapest tomatoes in area supermarkets at the time, but less expensive than on-vine greenhouse tomatoes and plastic boxes of cherry or grape tomatoes. But the real action went down behind the scenes off the back of the truck, where Stark and Miller, working from memory and a hand-scribbled list, personally filled the advance orders of their restaurant customers, who account for 60 to 70 percent of Eckerton's business. A parade of sous chefs from New York's premier eateries—Jean Georges, Gramercy Tavern, Eleven Madison Park, Babbo, Blue Hill, Daniel, Four Seasons, Telepan, Savoy, and Otto, to name a few—pushed carts or hailed taxis from behind stacks of produce flats beside the curb. Stark first-named his customers, engaging them in steady, lighthearted banter and jokes.

Peter Hoffman, the celebrated chef/owner of the restaurants Savoy and Back Forty in Manhattan, drifted by, wearing a pair of denim shorts and an old shirt. Even before it became a foodie gospel and bankable trend, Hoffman focused on buying local products and was a driving force behind the creation of Chefs Collaborative, an organization of culinary professionals dedicated to working closely with local farmers. Consulting a sheet of creased paper, he said, "One large and one medium," chef shorthand that meant that he had come to pick up one flat of assorted large heirloom tomatoes and one of medium. Stark handed him the flats, and Hoffman asked whether he could leave them on the sidewalk beside the truck and pick them up after he had finished trawling through the rest of the market. I took the opportunity to ask him why he and other chefs flocked to Stark's stand. He paused to think and then replied matter-of-factly, "It's all about taste, really."

Stark has built his business on the ineffable flavor of a real tomato, the very trait that industrial tomato producers have bred out of their product in a rush toward higher yields, disease resistance, toughness, shelf life, and round uniformity—"something red to put in their salad."

Hoffman said, "He's taken a lot of time and care in the field to make his soil healthy and that results in deeper flavors. He picks when his tomatoes are ripe. And he has passion and devotion to doing a good job, and ultimately we can taste that." If you were to order a salad of Stark tomatoes in one of Hoffman's restaurants, you might be surprised how little of the chef's hand is reflected in the dish compared to the farmer's. "Basically I cut the tomatoes and arrange a beautiful plate with all the different sizes and shapes. Then I drizzle it with some good oil and an acid and some great salt—maybe some fresh herbs," Hoffman said, gesturing toward the throngs of ordinary shoppers who swarmed in the spaces between Greenmarket's stalls. "Farmers' markets are burgeoning. Thousands and thousands of people shop in them every day. The lesson is that people really appreciate good flavor. You can fool a lot of folks into eating crap, but they notice the difference immediately when you give them something truly good. That's what Tim has done."

By noon, Eckerton's truck was almost empty. Stark grabbed a Brandywine out of a box and walked across the street to a falafel vendor. He gave it to the puzzled Middle Eastern cook, who had a full stainless steel pan of chopped food-service tomatoes on the counter. "Use this one in ours," Stark said, explaining that he had a stand at the market and had grown the tomato himself. Lunch in hand, we climbed back aboard Stark's pickup for the return drive to Pennsylvania. But his workday was far from over. Before we had made it across town to the tunnel, he phoned back to an intern at Union Square to remind her to collect an overdue payment from a restaurant. In a frantic call that came in as we rolled through Newark, New Jersey, Miller, who was still working the back of the box truck, announced that they had nearly sold out. There wouldn't be enough beefsteak tomatoes to fill a standing order from Tom Colicchio's 'wichcraft—an upscale sandwich and takeout place that exclusively features Stark's tomatoes on its signature BLT. Once Eckerton runs out of beefsteaks for the season, the BLT is removed from the menu until the following summer. And Stark had a crop of late tomatoes that he had hoped to keep selling to 'wichcraft for at least a few more weeks. "I try to spread out what I have with all the people I deal with. You don't want the chef from ESCA coming and asking how come he didn't get any tomatoes when one of his competitors did," he said. Steering through the traffic, Stark reached the cell phone of Nacho, who as usual had been left in charge of activities back on the farm. Were there any ripe beefsteaks? Could they be picked and packed immediately? Was anyone available to make a special trip into the city to keep an important customer supplied? Then he was on the phone to a chef at 'wichcraft. When, exactly, did he need beefsteaks? Would tomorrow do, or did it have to be today? As soon as Stark hung up, his phone rang again. It was Nacho. Yes, ripe beefsteaks were available. Yes, someone could drive them into town. Another crisis averted.

Thinking again about what Stark had said about niche marketing, I was reminded of a passage in *Omnivore's Dilemma*, in which Michael Pollan discusses "artisanal economics," a theory that Allan Nation outlined in the magazine *Stockman Grass Farmer*. Nation contends that industrial farmers sell commodities, crops that are intentionally produced to be identical to each other. The only way to compete, according to Nation, is to offer goods that cost less than the next farmer's. A rush to the bottom becomes inevitable.

Clearly, that is what has happened to commercial growers in Florida, who struggle to compete with nearly identical tomatoes grown in

Mexico and with hydroponic produce from Canada. Artisanal economics—of which Eckerton's efforts are a perfect example—turns that approach on its head. It celebrates oddness. Uniformity is odious, variation sought after. Stark's competitive advantage comes from being special and selling an exceptional product to a local market, where free word of mouth replaces expensive advertising campaigns. Instead of trying to fix the "bad" qualities of a tomato—softness, differing shapes and sizes, a restricted growing season—Stark embraces the fruit's intrinsic "tomato-ness" and in doing so has built a business that allows his employees to buy cars, purchase homes, and send children to private high schools back in Mexico. He doesn't harm the land or sicken his workers with chemicals. And he and other farmers like him have put good-tasting tomatoes that customers can feel good about buying within reach of every person living near a farmers' market.

As we crossed the Delaware River back into Pennsylvania at about two o'clock in the afternoon, Stark got a call from Miller at Greenmarket. They were packing up early. Every last tomato—two tons in total—had been sold. Eckerton had grossed nearly $15,000. Payroll would be made again. For the first time in thirty hours, Stark relaxed, exhaling loudly. "Maybe I'll get tired of this someday," he said. "But for me, for now, it seems like the right thing to do."

WILD THINGS

I went to northwestern Peru hoping to find a wild *Solanum pimpinellifolium*, the progenitor of the tomatoes we eat. But after an hour's drive north of the regional capital of Trujillo on the Pan American Highway, I figured I would be fortunate to encounter any living plant. The desert stretched away on both sides of the road. From the hazy peaks of the Andes lying to the east, to a gray wall of fog stretching west over the Pacific Ocean, the landscape was devoid of life—not a tree, bush, blade of grass, or cactus to be seen. In comparison, the Sonoran and Mojave Deserts of California and Arizona look like verdant pastures. Roger Chetelat, the tomato geneticist at the University of California Davis, had given me the geographical coordinates of the sites where his research team had spotted wild tomato populations while doing field research in 2009. He had also warned me that my chances of finding one were not great, because of the pressures of urbanization and large-scale agriculture. He had e-mailed a list of local names for *S. pimpinellifolium*: *tomatito* (little tomato), *tomate cimarrón* (wild tomato), *tomate del campo* (field tomato), *tomate de culebras* (snake tomato), *tomate de zorro* (fox tomato), and *tomates silvestres* (wild tomatoes) to share with my guides. Before we left town, my driver, Carlos Chavez Garcia, read it over, shrugged, and passed it to another driver, who also shrugged. Neither of them recognized any of them.

Chavez and I had not seen a single tomato when he turned his mint-condition twelve-year-old Toyota Corolla off the Pan American Highway and began to negotiate a twisty secondary road that paralleled the Rio Jequetepeque into the Andes. Fed by melting snows from the mountains, rivers such as the Jequetepeque were diverted by the region's earliest human inhabitants into a network of canals that allowed farming to flourish and civilizations to develop. Some of those same canals are still in use today. Chavez drove past fields of rice, corn, leafy greens, and tomatoes of the domestic variety. But not a single *S. pimpinellifolium*. As Chetelat had warned, every arable patch of land was tilled, right to the banks of the canals and the edges of the road.

As we rose higher, the farms dwindled, until eventually we were back in a lifeless land of sheer cliffs and boulder-strewn slopes, surrounded on all sides by jagged, barren peaks. Chetelat had given me the coordinates for "a pretty good cluster" just outside Tembladera, a neat little town on the banks of a turquoise-colored reservoir. When my handheld GPS receiver ushered us to the spot he had described, there were no plants, just a steep, rocky valley. Chavez stopped the car and consulted three women who were walking away from town carrying plastic shopping bags. They chattered, gesturing up the valley, and shook their heads. Chavez came back to the car. "Wrong time of year," he said. "They told me that there were wild tomatoes here in the summer, but not now." He went on to say that the plants I was looking for were called *tomatillos silvestres* by locals and that he remembered snacking on them as a boy on his grandmother's farm outside of Trujillo, but that he hadn't seen any in years. "They are gone," he said.

Chavez pulled a U-turn and we began to drive back toward the highway. The car hadn't traveled fifty yards when I caught a flash of yellow. "Stop!" I cried. Chavez had to continue some distance along the road before the shoulder widened enough for him to get the car halfway off the pavement. He eyed me with skepticism. I trotted back toward where I had seen the flowers. There, growing out of the base of what seemed like a solid rock ledge without a trace of earth was a sprawling, jagged-leafed vine. Easily recognizable as a member of the tomato clan, it was covered in sunny yellow flowers, tiny green fruits, and near its base, bright, red miniature tomatoes not much bigger than cranberries. Catching up to me, Chavez said delightedly, "*Tomatillos silvestres*!" and began picking them off the vine and popping them into his mouth as fast as he could, pausing occasionally to repeat, "*Tomatillos silvestres*!"

I picked one for myself and brushed off the road dust with my shirttail. The fruit between my thumb and index finger was as smooth and spherical as a marble. I gave it a squeeze, and it did not yield. I threw it down onto the pavement to see what would happen. It was undamaged, and I popped it into my mouth. The bright, sweet pop of taste was followed by a lingering, pleasant tartness—that essential balance that defines a great tomato.

Afterword

NEW WORLD, OLD CHALLENGES

On a crisp, sunny spring morning in 2011, I did something that could well have gotten me arrested only a few months earlier. I rode south of Immokalee in a road-weary Toyota with Lucas Benitez and three other members of the Coalition of Immokalee Workers (CIW). We slowed as we arrived at the security gates in front of a packing plant owned by Pacific Tomato Growers, a major fruit and vegetable producer with farms in Florida, Georgia, Virginia, California, and Mexico. Consumers know the company through its brand names Sunripe and Suncoast. During the course of nearly two decades of struggling to improve working conditions and end abuses in Florida's fields, the CIW had come calling on the Pacific facility numerous times to present its demands, each time to be met with locked gates. "One time, there were sheriff's officers with guns," said Benitez. On this occasion, officials at the company did not know that I was accompanying the CIW members. I scrunched in my seat, trying to look inconspicuous.

I needn't have bothered; the watchman gave a smile and a nod as the car cruised past the guardhouse. In a parking lot beyond it, we encountered a gathering that would have been as unlikely at an earlier time in the Florida tomato patch as the CIW's visit to the packing plant. A school bus had stopped in front of a building that resembled a carport. Its wooden trusses and timbers were fresh, blond colored, and had obviously been removed only recently from a lumber warehouse, nailed together, and exposed to the sun and humidity of Florida. A bank of solar panels rested in its roof. Members of a harvest crew filed from the bus and stood in front of what looked like a bank of ATMs. Actually, the machines were yet another novelty in the Florida tomato industry: punch clocks that would keep track of every minute the pickers worked and guarantee that they received the right wages.

Once on the clock and being paid for their time, the workers—Hispanic and clad in jeans, T-shirts, and either baseball caps or straw cowboy hats—ambled in groups of two or three toward a modular building with a sign near the door that read TRAINING. About fifty pickers had already gathered on folding chairs inside when the CIW delegation entered and began preparing to lead a class designed to educate the field crew about their rights and responsibilities under a new program called the Fair Food Agreement. I took a seat in the far back corner.

The Fair Food Agreement had been signed only five months earlier on a folding table under a tree in back of the CIW headquarters by Reggie Brown, the executive vice president of the Florida Tomato Growers Exchange, the agricultural cooperative whose members grow virtually all of Florida's tomatoes. According to the terms of the Fair Food Agreement, the growers promised to pass along an extra penny a pound to pickers if—and this is a crucial "if"—the end buyer of the tomatoes had also signed the agreement and agreed to pay the penny. It doesn't sound like much, but for an average worker it's the difference between making $50 and $80 a day—between abject poverty and a living, albeit hardscrabble, wage. More important, the agreement also requires that companies use time clocks like the one the workers were punching that morning to make sure that they earn at least the $7.25 minimum hourly wage to which they are entitled. A system has been put in place so workers can complain about injustices without fear of being fired. Health and safety rules were established, requiring farms to provide shade for workers; guaranteeing them a sufficient number of work breaks, including time for lunch; and introducing a stern policy to combat sexual harassment—all too common in the fields. These are all rights that the average American takes for granted, but they had never applied to tomato workers. It was as if a Dickensian workhouse had overnight adopted the labor practices of a state-of-the-art auto plant.

But shaking hands and signing a feel-good agreement in the shade of a tree is one thing. Making that agreement work in gritty, remote Florida tomato fields is an entirely different matter. The coalition members and the executives of the companies represented by Reggie Brown, who had previously not even been on speaking terms, suddenly found themselves in a position where they would have to communicate and work cooperatively to transform the fine words of the Fair Food Agreement into tangible deeds. Together, the growers' and workers' representatives decided that the 2010-to-2011 growing season would be a period of transition. Pacific and Six L's (since renamed Lipman), the two large tomato growers on whose fields pickers who worked under the infamous Navarrete slavery gang had once toiled, took the lead. Almost immediately after the agreement was signed, they began to work with the coalition to find practical ways of introducing a new Code of Conduct to the Florida tomato industry on a limited basis. Once they had worked through practical kinks at their two companies, the policies would be rolled out during the 2011-to-2012 growing season to every member of the exchange that had signed the Fair Food Agreement—meaning that virtually every tomato picker in Florida would benefit from higher wages; more stringent safety standards; a clear system for lodging complaints; and protection from sexual harassment, enslavement, and other abuses. To make sure that they understand their rights and responsibilities under the new system, all fifty thousand workers who pick tomatoes in Florida each year will attend classes like the one I joined last spring. They also will be required to watch a video that illustrates those rights in simple, clearly comprehensible images and language.

Inside the education building, the CIW's Benitez was greeted by Angel Garcia, the human-resources manager of Pacific. Physically, the two men were a study in contrasts: Benitez, short and compact, with a wrestler's build and an intense demeanor; Garcia, large, bearlike, and soft spoken. In the 1990s, Benitez was just another Mexican teenager loading tomatoes into a bucket in the fields spreading out to the horizon behind the packing house. Today, he and Garcia would address members of a crew much like the one to which Benitez once belonged.

Over the next forty minutes, Benitez and the three other coalition members spoke to the workers. In Spanish, they explained that the minimum wage in Florida was $7.25 per hour, and that the company was required to have time clocks to keep track of the hours worked. "The $7.25 is the minimum," Benitez said. "If you're waiting for dew to dry, it counts as time worked. The same if a truck has not arrived, or if it begins to rain and you have to stop picking but are not taken home." He stressed that the $7.25 minimum applied even if they were working by the piece (paid per bucket). If they picked enough buckets to exceed $7.25 per hour, that was a bonus for their hard work.

Benitez then lifted one of the containers that workers use to collect tomatoes onto the table in front of him. Called *cubetas*, the flower pot–shaped buckets are meant to hold thirty-two pounds of slicing tomatoes when properly filled. The *cubeta* that Benitez had put on the table was piled high with bright green tomatoes that were mounded above the rim of the container like ice cream scooped on top of a cone, the way full *cubetas* carried by pickers usually look. Benitez swept his arm over the top of the *cubeta* in the manner of someone leveling a cup of flour. In the process, he knocked some of the heaped tomatoes onto the table. "This is what a full *cubeta* is supposed to look like," he said, emphatically. "There should be no tomatoes in it that are completely above the rim." The workers stood, peered, and began talking animatedly among themselves.

I didn't understand the intensity of the pickers' interest until later that day, when Benitez explained to me that the definition of what constituted a full bucket had long been a source of tension, disputes, and occasionally fistfights in the fields. Workers, he said, run with the filled buckets to a truck and throw them up to a low-level field manager called a *dompeador*, who takes the bucket, dumps the tomatoes into a bin, and returns the bucket empty, along with a ticket that the worker pockets. At the end of the day, pickers count their tickets and are paid accordingly. But it was common for bosses to insist that workers overfill their buckets. The CIW calculated that mounding tomatoes resulted in workers picking about 10 percent more tomatoes than they were paid for—the difference going to the leader of the field crew and the company. Often a *dompeador* would refuse to give a picker a ticket for a bucket he deemed insufficiently full, or would send the worker back to the field to top one up. "It was a constant source of friction and humiliation for the workers," said Benitez. It was also one of the problems that the CIW and growers had to work out. To do so, they met in the CIW offices with an empty *cubeta*, a pile of tomatoes, and a scale. By weighing out exactly thirty-two pounds, they agreed on precisely what a full *cubeta* should look like.

The coalition members also outlined safety policies introduced under the Fair Food Agreement and told the workers that there would be zero tolerance of sexual harassment—no dirty jokes, unwelcome touching, demands for sex, promise of job benefits for sex, or sexual assault of any kind.

At the end of the meeting, Garcia, Pacific's HR boss, stood up. The room went quiet, and I wondered what he would say. Would he remind the pickers that they worked for him, not the CIW? Would he insist that if they did have a problem to report, that they first approach officials within the company? Would he give a pep talk urging them to get out into the fields and bring in the day's harvest?

Instead, he dealt out a wallet card to each crew member, saying that the cards bore his direct phone number, as well as the number of a 24-hour employee hotline. Then he began to speak, slowly and softly. "Call us if you need help," he said. "It is confidential." He said that the company did not want to have the abuses of the past repeated. Then he said, "If you see something, talk to somebody. Tell the guy next to you. Tell your boss. Tell me. Tell the CIW. Tell anybody; but say something."

As the workers filed out of the classroom on their way to a day of picking grape tomatoes, the CIW members handed each a booklet. Printed on thick glossy paper and illustrated with color photographs, in twelve pages of simple Spanish, it outlined the new rights that the coalition had worked for nearly two decades to bring to the fields.

Despite the successes of the CIW, there is still one major impediment to tomato pickers fully enjoying the benefits of the newly won rights and wage increase. Although McDonald's, Burger King, and most other fast-food and food-service corporations have signed the Fair Food Agreement, the nation's supermarket chains, with the notable exception of Whole Foods Market, refuse to deal with the CIW. This not only prevents the workers from getting the full benefit of the "penny-per-pound" deal for tomatoes sold to supermarkets (grocery store chains buy about half of Florida's tomatoes), but can leave entire sectors of the workforce without the protections of the moral and financial clout of major buyers pledging not to deal with growers who willfully violate the Code of Conduct. As one CIW member described it, the coalition has built the conduit through which rights and money can flow. Now it's up to the supermarket industry to fill that conduit, if not on its own volition, then with a little of the encouragement the CIW is so effective at giving.

My protest-marching skills had grown a bit rusty, having last been put to use in 1968, on behalf of Eugene McCarthy and his thwarted bid for the Democratic presidential nomination. But on a bitter afternoon in Boston in 2011, I sloshed through a few inches of slushy snow with more than nine hundred supporters of the two busloads of CIW members who had traveled from Florida, sleeping on church floors and bathing at homeless shelters along the way. We tramped along Huntington Avenue, a broad four-lane boulevard, from Boston's Copley Square to a Stop & Shop supermarket, part of a northeastern chain owned by Ahold, a Netherlands-based multinational grocery chain. The workers wanted Stop & Shop to sign a Fair Food agreement. With a rambunctious brass band, clever signage, and rousing warm-up speeches by *Diet for a Small Planet* author Frances Moore Lappé, Slow Food USA president Josh Viertel, and Lucas Benitez, it was a friendly, festive event. ("We like to have a bit of fun at our actions," Gerardo Reyes of the CIW told me.) But its purpose could not have been more serious. On the day we marched, fresh slicing tomatoes were on sale for $1.99 a pound in Boston stores. Paying one penny more per pound for an out-of-season luxury that no one really needs seems a modest sacrifice that a $40-billion-a-year conglomerate like Ahold could easily absorb, if it wished to. When the throng of marchers arrived at the Stop & Shop, a manager came out and accepted a letter from the CIW, but Ahold gave no indication of interest in signing the Fair Food Agreement—at least on that day. But there will be others. "We didn't come this far by fainting if companies didn't come to the table the first time we called," said Greg Asbed of the coalition.

Although the CIW has spared no large supermarket chain in its drive for more signatories to the Fair Food Agreement, it has turned its focus to two companies in particular. Publix Super Markets, a large southeastern chain based in Florida, has been the target of dozens of actions—marches to company headquarters, pray-ins by religious leaders in its produce sections, demonstrators at every new store opening—but the company adamantly refuses to take part in the Fair Food Agreement. Initially, Publix said that it did not want to intervene between its suppliers and their workforce. But when the growers and the CIW began working as partners to implement the Fair Food Agreement in late 2010, Publix changed its tune, arguing that the extra penny a pound should be included in the price it pays growers for tomatoes. "It is not our place to enter into direct wage negotiations with employees of our suppliers," a spokeswoman for the company told the *Boston Globe*. "We will pay market price to suppliers who comply to our standards."

Of all the supermarkets approached by the CIW to take part in the Fair Food Agreement, the most perplexing reaction has come from Trader Joe's, a 360-plus-store chain that brands itself as a worker- and customer-friendly bastion of all things sustainable, organic, and fair trade. Its main competitor in the trendy grocery store niche, Whole Foods Market, came aboard early and willingly. Trader Joe's executives have been contacted, coalition supporters have marched on stores, and petitions have been sent, all to no avail. The chain, owned through a trust set up by the founder of Aldi, a Walmart-like German discount chain, contends that it more than meets the coalition's demands through its own purchasing policies.

In late 2011, about 400 workers' rights advocates marched to Trader Joe's headquarters outside Los Angeles. They wanted to present management with two letters, one signed by 109 rabbis, the other from more than 80 Southern California pastors. Both letters simply asked Trader Joe's executives to work with the coalition to address labor abuses in the tomato fields—nothing more. Noting that the story of their own religion began with the "journey of our ancestors from slavery to freedom," the rabbi's letter said, "this legacy informs out moral imperative to fight modern slavery and uphold the right of every individual to be free."

The religious leaders were disappointed when they were greeted at the locked doors to Trader Joe's headquarters by a uniformed security guard who said that no one at the company would accept the letters, and that he didn't even know the name of a soul who worked in the building. "I do a lot of workers' justice actions, and

I've never experienced a reaction like that," Rev. Sarah Halverson, of Fairview Community Church in Costa Mesa, California, told me. "Even Walmart always sends someone out to accept our letters. I didn't expect the president of the company to greet us, but they could have at least sent an assistant from the PR department." Halverson said that she peered through the glass and saw workers in the lobby wearing Trader Joe's iconic Hawaiian shirts. When they saw her, they scurried out of sight. "It's hard to believe that a group of rabbis and pastors was threatening," she said.

That snub was nothing compared with what happened next. Police arrived and ordered the interfaith leaders to disperse, which they did. But as Rev. Halverson walked away, she saw someone from inside the building open the door, step outside, rip the letters from the glass, and crumple them. "They crossed a line," she said. "At first I felt hurt, almost heartbroken, then angry. I was under the impression that they were a socially conscious store. Most members of my congregation are progressive, and that's why they shop there. But I lost complete respect for them. They showed their true colors. But as a pastor, I believe everybody has good in them, and that ultimately Trader Joe's will sign the Fair Food Agreement."

Earlier, at the Boston rally, CIW member Lucas Benitez told the marchers, "This greed has to change. The executives at grocery chains want what is best for their children; we want the same for ours. We will not stop." Supermarket operators should heed those last four words. From the moment of its birth two decades ago, when a handful of migrant workers gathered in a church meeting room in Immokalee, through its early efforts to stop beatings in the fields and collect wages owed by recalcitrant crew bosses, to demonstrations, hunger strikes, boycotts, shareholder actions, pray-ins, petitions, letter-writing campaigns, undercover operations to free enslaved comrades, and testimony before the Senate, stopping is the one thing that the CIW has never done.

Notes

INTRODUCTION: ON THE TOMATO TRAIL

xiv If you have ever eaten: Statistics from the Florida Tomato Commission's *Tomato 101*, http://www.floridatomatoes.org/facts.html.

xiv Americans bought $5 billion: U.S. Department of Agriculture Economics, Statistics, and Market Information System, *U.S. Tomato Statistics*, Table 070 and 076. I multiplied the average retail price in 2009 by the total production.

xiv In survey after survey: See Christine M. Bruhn, Nancy Feldman, Carol Garlitz, Janice Harwood, Ernestine Ivans, Mary Marshall, Audrey Riley, Dorothy Thurber, Eunice Williamson, "Consumer Perceptions of Quality: Apricots, Cantaloupes, Peaches, Pears, Strawberries, and Tomatoes," *Journal of Food Quality* vol. 14, no. 3 (July 1991): pp. 187–95.

xiv According to analyses: Thomas F. Pawlick, author of *The End of Food: How the Food Industry Is Destroying Our Food Supply—And What You Can Do About It* (Fort Lee, NJ: Barricade Books, 2006), originally presented this information. I have updated it. The 1960s figures come from Bernice K. Watt and Annabel L. Merrill, *Composition of Foods: Raw, Processed, Prepared*, U.S. Department of Agriculture, Agricultural Research Service, Agricultural Handbook No. 8 (Washington, DC, 1964). My source for 2010 figures is the U.S. Department of Agriculture Nutrient Database for Standard Reference, Release 23: http://www.ars.usda.gov/SP2UserFiles/Place/12354500/Data/SR23/sr23_doc.pdf.

xv A couple of winters ago: Pawlick (see above) performed a similar "experiment."

xv Little wonder that tomatoes are by far the most popular: National Gardening Association, "The Impact of Home and Community Gardening in America" (2009), http://www.gardenresearch.com/index.php?q=show&id=3126.

xvi Regulations actually prohibit: Federal Marketing Order No. 966 sets standards for tomatoes exported from most of Florida during the colder months.

xvii To get a successful crop: Stephen M. Olson and Bielinski Santos. eds., *Vegetable Production Handbook for Florida 2010–2011*, University of Florida (2010): pp. 295–316.

xvii Not all the chemicals stay behind: The source is the Environmental Working Group, which compiled statistics from the U.S. Department of Agriculture's Pesticide Data Program and the U.S. Food and Drug Administration's Pesticide Monitoring Database.

xviii The industry was nearly dealt: For salmonella losses, see Mickie Anderson, "UF Research Finds Salmonella Responds Differently to Varieties, Ripeness," *University of Florida News*, September 21, 2010. For freeze losses see Laura Layden, "Florida Tomato Growers Eye Rebound from 2009–2010 Freeze-Ravaged Season," *Naples Daily News*, October 3, 2010. For the effects of glut, see Liam Pleven and Carolyn Cui, "Dying on the Vine: Tomato Prices—Tomatoes Go from Shortage to Glut in a Matter of Weeks," *Wall Street Journal*, June 17, 2010.

xviii This has put a steady downward pressure: Source for wage statistics is the Coalition of Immokalee Workers, http://www.ciw-online.org/Resources/10FactsFigures.pdf.

xix The owners had crop insurance: Michael Peltier, "The Other Side of the Freeze," *Naples Daily News*, February 8, 2010.

xix And conditions are even worse: See "From the Hands of a Slave" in this book.

xx Labor protections for workers predate the Great Depression: Farmworkers were specifically exempted from the Fair Labor Standards Act of 1938, a key component of Franklin D. Roosevelt's New Deal.

ROOTS

1 A Chilean Soldier was guarding: A January 7, 2010, interview with Roger Chetelat in his office at the University of California Davis provided much of the information on the Atacama Desert expedition and tomato genetics. For a more scientific description, see Roger T. Chetelat, Ricardo A Pertuzé, Luis Faúndez, Elaine B. Graham, and Carl M. Jones, "Distribution, Ecology and Reproductive Biology of Wild Tomatoes and Related Nightshades from the Atacama Desert Region of Northern Chile," *Euphytica* vol. 166 (December 25, 2008): pp. 77–93.

2 The Atacama Desert makes up: See Yuling Bai and Pim Lindhout, "Domestication and Breeding of Tomatoes: What Have We Gained and What Can We Gain in the Future?," *Annals of Botany* vol. 100, issue 5 (August 23, 2007): pp. 1085–1094.

3 one of our favorite vegetables: Hayley Boriss and Henrich Brunke, "Commodity Profile: Tomatoes Fresh Market," University of California, Agricultural Marketing Resource Center (October 2005).

4 When Hernán Cortés conquered: For the history of the tomato, I drew on Andrew F. Smith, *The Tomato in America: Early History, Culture, and Cookery* (Columbia: University of South Carolina Press, 1994); and Arthur Allen, *Ripe: The Search for the Perfect Tomato* (Berkeley, CA: Counterpoint, 2010).

6 Tomatoes' near-universal popularity: A. W. Livingston, *Livingston and the Tomato*, forward and appendix by Andrew F. Smith (Columbus: Ohio State University Press, 1998). The autobiography of the great early plant breeder benefits enormously from Smith's writing and scholarship.

6 **"Well do I remember":** ibid p. 19

8 **Florida was a late comer:** For reference to Parry, Wilson, and Blund, see S. Bloem and R. F. Mizell, "Tomato IPM in Florida," University of Florida, Institute of Food and Agricultural Sciences Extension, Publication no. ENY706/IN178, http://edis.ifas.ufl.edu/in178. For Hendrix, see Benjamin Bahk and Mark Kehoe, "A Survey of Outflow Water Quality from Detention Ponds in Agriculture," Southwest Florida Water Management District (1977) and http://floridahistory.org/palmetto.htm.

9 **That was around the time**: See E. F. Kohman, "Ethylene Treatment of Tomatoes," *Industrial and Engineering Chemistry* (October 1931): pp. 1112–13.

9 **The Person most responsible:** Statistics from the U.S. Department of Agriculture Economics, Statistics, and Market Information System, Table 016, http://usda.mannlib.cornell.edu/MannUsda/viewDocumentInfo.do?documentID=1210.

10 **Max Lipman:** See Carlene A. Thissen, *Immokalee's Fields of Hope* (New York: iUniverse, 2004); also the Web site of Six L's Packing Company, http://www.sixls.com.

12 **Born in Reading:** I am deeply in debt for information about Charles Rick from Arthur Allen, *Ripe: The Search for the Perfect Tomato* (Berkeley, CA: Counterpoint, 2010), which contains an excellent minibiography of the legendary plant science professor. I also drew on an interview and profile written by Craig Canine, "A Matter of Taste: Who Killed the Flavor in America's Supermarket Tomatoes?" *Eating Well* (January/February 1991): pp. 40–55.

A TOMATO GROWS IN FLORIDA

19 **When I met Monica Ozores-Hampton:** Details about commercial tomato horticulture in Florida in this chapter came from an interview with Ozores-Hampton on June 2, 2010. Any errors are my own. Information about the possible health effects about pesticides was taken from reports of the Pesticide Action Network and in no way reflects Ozores-Hampton's opinions.

24 **If those roots:** The Pesticide Action Network's database on methyl bromide can be accessed at http://www.pesticideinfo.org/Detail_Chemical.jsp?Rec_Id=PC32864.

26 **more than one hundred chemicals:** See Stephen M. Olson and Bielinski Santos, eds., *Vegetable Production Handbook for Florida 2010–2011*, University of Florida Institute of Food and Agricultural Sciences (2010), pp. 295–316.

26 **Six of the recommended herbicides:** The Pesticide Action Network's database for agricultural chemicals can be found at http://www.pesticideinfo.org/Search_Chemicals.jsp#ChemSearch.

27 A distressing number: The Environmental Working Group compiled statistics from the U.S. Department of Agriculture's Pesticide Data Program and the U.S. Food and Drug Administration's Pesticide Monitoring Database. See also Thomas J. Stevens III and Richard L. Kilmer, "A Descriptive and Comparative Analysis of Pesticide Residues Found in Florida Tomatoes and Strawberries," University of Florida Institute of Food and Agricultural Sciences, 1999.

28 Joseph Procacci agreed to take me: The Procacci interview took place on March 2, 2005.

32 To see the next phase: Steven A. Sargent, Jeffrey K. Brecht, and Teresa Olczyk, "Handling Florida Vegetables Series: Round and Roma Tomato Types," University of Florida Institute of Food and Agricultural Sciences, 1989, gives a good overview of postharvest tomato packing.

32 rise to 110 degrees: Jeffrey K. Brecht, a postharvest physiologist at the University of Florida Research Center, made this statement at a workshop for packinghouse managers in 2006: http://www.gladescropcare.com/GCC_TPHMW.pdf.

32 Despite such sanitation: See *Program Information Manual: Retail Food Protection Storage and Handling of Tomatoes*, U.S. Food and Drug Administration (June 10, 2010), http://www.fda.gov/food/foodsafety/retailfoodprotection/industryandregulatoryassistanceandtrainingresources/ucm113843.htm; and Martha Roberts, Florida Tomato Committee, (in an address to the 2006 Florida Tomato Institute, http://www.gladescropcare.com/GCC_TPHMW.pdf).

CHEMICAL WARFARE

35 Tower Cabins is a labor camp: "Why Was Carlitos Born This Way?"—the story of the birth defects in Immokalee—was broken on March 13, 2005, in the *Palm Beach Post* by reporter John Lantigua. I am in debt to Lantigua and his colleagues Christine Stapleton and Christine Evans for many of the details of this tragedy, which might never have come to light had it not been for their doggedness and insightfulness.

35 But in the lives of tomato workers: Geoffrey M. Calvert, Walter A. Alarcon, Ann Chelminski, Mark S. Crowley, Rosanna Barrett, Adolfo Correa, Sheila Higgins, Hugo L. Leon, Jane Correia, Alan Becker, Ruth M. Allen, and Elizabeth Evans, "Case Report: Three Farmworkers Give Birth to Infants with Birth Defects Closely Grouped in Time and Place—Florida and North Carolina, 2004–2005," *Environmental Health Perspectives* vol. 115, no. 5 (May 2007): pp. 787–91.

36 Many of them were rated "highly toxic": The Pesticide Action Network's database for agricultural chemicals is http://www.pesticideinfo.org/Search_Chemicals.jsp#ChemSearch.

36 "restricted entry intervals": For a list of pesticides used on tomatoes in Florida and their restricted entry intervals, see "Florida Crop/Pest Management Profiles: Tomatoes," University of Florida, Institute of Food and Agricultural Sciences (March 2009). http://edis.ifas.ufl.edu/pi039.

37 Although regulations require: These regulations vary depending on which pesticide is used. For the U.S. Environmental Protection Agency's Worker Protection Standard for Agricultural Pesticides, see http://www.epa.gov/oecaagct/twor.html.

38 As soon as I met him: Much of the background material in this chapter came from a June 2, 2010, interview with Andrew Yaffa.

41 In terms of raw quantities: "Agricultural Chemical Usage 2006 Vegetable Summary," U.S. Department of Agriculture, National Agricultural Statistics Service (July 2007).

41 Employing only about fifty inspectors: "Abundance of Poisons, Shortage of Monitoring," *Palm Beach Post*, May, 1, 2005.

41 workforce of roughly 400,000: "National Agricultural Workers Survey," U.S. Department of Labor Employment and Training Administration, October 5, 2010 (Washington, D.C.).

42 Together for Agricultural Safety: See Joan Flocks, Leslie Clarke, Stan Albrecht, Carol Bryant, Paul Monaghan, and Holly Baker, "Implementing a Community-Based Social Marketing Project to Improve Agricultural Worker Health," *Environmental Health Perspectives Supplements* vol. 109, no. S3 (June 2001): pp. 461–688.

42 less than 8 percent: John Lantigua, "Why Was Carlitos Born This Way?" *Palm Beach Post*, March 13, 2005.

42 leveled eighty-eight counts: Laura Layden, "Judge: Drop Most Violations against Ag-Mart," *Naples Daily News*, March 23, 2007.

43 A scathing portrait: Shelly Davis and Rebecca Schleifer, "Indifference to Safety: Florida's Investigation into Pesticide Poisoning of Farmworkers," Farmworker Justice (1998), Washington, DC, http://www.farmworkerjustice.org/pesticides/173-indifference-to-safety.

43 agricultural workers are more likely to be poisoned: See *Worker Health Chartbook, 2004*, U.S. Department of Health and Human Services, Centers for Disease Control and Prevention, National Institution for Occupational Safety and Health, p. 138. http://www.cdc.gov/niosh/docs/2004-146/pdfs/2004-146.pdf.

44 Guadalupe Gonzales III: This information came from "Pesticide Use Inspection Report," file no. 101-266-4076, "Gonzales III, Guadalupe," acquired through a public records request to the Florida Department of Agriculture and Consumer Affairs.

46 virtually no hard scientific research: See "Improvements Needed to Ensure the Safety of Farmworkers and Their Children," U.S. General Accounting Office (March 2000), http://www.gao.gov/new.items/rc00040.pdf.

47 Leaning on her cane: I interviewed Linda Lee and Jeannie Economos on June 3, 2010.

47 In a survey of workers conducted: Ron Habin, "Lake Apopka Farmworkers Environmental Health Project Report on Community Health Survey," Farmworker Association of Florida (May 2006).

48 Located fifteen miles northwest of Orlando: "Lake Apopka Timeline," Friends of Lake Apopka, http://www.fola.org/PDFs/LakeApopkaTimeline.pdf.

48 In one sweet deal: Edward Ericson Jr., "A Cool Deal, Going Once, Going Twice," *Orlando Weekly*, December 12, 1998.

49 researchers determined that the cause of the deaths was pesticide poisoning: See "Final Lake Apopka Natural Resource Damage Assessment and Restoration Plan," United States Fish and Wildlife Service (June 2004). http://restoration.doi.gov/Case_Docs/Restoration_Docs/plans/FL_Lake_Apopka_RP_06-04.pdf.

50 One Sunday morning: Sara Olkon, "Pesticide Drift to Be Investigated: Churches Fear Effect of Toxin," *Miami Herald*, February 22, 2001.

50 Subsequent air tests: Described in a joint press release from the Farmworker Association of Florida and the Friends of the Earth (February 22, 2001).

51 initiated thirty-nine investigations: Frederick M. Fishel and J. A. Ferrell, "Managing Pesticide Drift," The University of Florida, Institute of Food and Agricultural Sciences, http://edis.ifas.ufl.edu/pi232.

51 "This will kill agriculture": This and the quotation later in the paragraph come from an article by Richard Dymond, "Growers Don't Like the Smell of Zone Bill," *Bradenton Herald*, June 9, 2007.

52 According to a 2009 report: Mark Mossierm, Michael J. Aerts, and O. Norman Nesheim, "Florida Crop/Pest Management Profiles: Tomatoes," University of Florida, Institute of Food and Agricultural Sciences, CIR 1238 (revised March 2009).

52 **a team led by Karen Klonsky:** K. M. Klonsky and R. L. De Moura, "Sample Costs to Produce Processing Tomatoes," University of California Davis, Cooperative Extension (2001); and "Production Practices and Sample Costs for Organic Processing Tomatoes in the Sacramento Valley," University of California Davis, Cooperative Extension (1993–1994).

52 **In Florida, nematodes:** Stephen M. Olson and Bielinski Santos, eds., *Vegetable Production Handbook for Florida 2010–2011*, University of Florida Institute of Food and Agricultural Sciences (2010): pp. 29–38, 47–54.

53 **the fumigant was approved:** See "Extension of Conditional Registration of Iodomethane (Methyl Iodide)," U.S. Environmental Protection Agency, August 13, 2009, http://www.epa.gov/pesticides/factsheets/iodomethane_fs.htm.

53 **despite a letter of warning:** The letter was sent by Robert G. Bergman of the University of California Berkeley and Ronald Hoffmann of Cornell University to Stephen Johnson at the U.S. Environmental Protection Agency on September 24, 2007.

53 **A report by the federal government:** This was from U.S. State Department document "USA CUN11 SOIL TOMATOES Open Field," a federal government application for an exemption to the methyl bromide ban on tomatoes submitted in 2009 for the year 2011.

53 **There are already signs:** Jacob Adelman, "Calif. Pesticide Opponents Deploy Florida Report," *San Jose Mercury News*, September 14, 2010.

55 **Dr. J. Routt Reigart:** From the February 28, 2008, deposition of John R. Reigart, Case No. 06-001725, Circuit Court of the Thirteenth Judicial District in and for Hillsborough County, Florida, Division B, *Francisca Herrera and Abraham Candelario v. Ag-Mart Produce, Inc.*

55 **Dr. Kenneth Rudo:** From the July 9, 2007, deposition of Kenneth Mark Rudo, Case No. 06-001725, Circuit Court of the Thirteenth Judicial District in and for Hillsborough County, Florida, Division B, *Francisca Herrera and Abraham Candelario v. Ag-Mart Produce, Inc.*

56 **On the morning of June 23, 2006:** From the June 23, 2006, deposition of Francisca Herrera, Case No. 06-001725, Circuit Court of the Thirteenth Judicial Circuit in and for Hillsborough County, Florida, Division B, Francisca Herrera and Abraham Candelario vs. Ag-Mart Produce, Inc.

59 **After a break:** From the June 23, 2006, deposition of Abraham Candelario Alphonso.

60 **Cisneros and I agreed:** Interviewed Cisneros on June 4, 2010. She also repeated much of what we spoke about during our lunch at her September 13, 2006, deposition. See above.

65 When word reached him: John Lantigua, "Produce Firm President Talks to Parents of Children with Defects," *Palm Beach Post*, March 26, 2006.

65 Yaffa's five-hour deposition: The deposition took place on August 22, 2006. See above.

71 To Yaffa's disappointment: Laura Layden and Janie Zeitlin, "Health Officials: Pesticides Not Likely at Fault for Birth Defects," *Naples Daily News*, October 13, 2005; and Geoffrey M. Calvert, Walter A. Alarcon, Ann Chelminski, Mark S. Crowley, Rosanna Barrett, Adolfo Correa, Sheila Higgins, Hugo L. Leon, Jane Correia, Alan Becker, Ruth M. Allen, and Elizabeth Evans, "Case Report: Three Farmworkers Give Birth to Infants with Birth Defects Closely Grouped in Time and Place—Florida and North Carolina, 2004–2005," *Environmental Health Perspectives* vol. 115, no. 5 (May 2007): pp. 787–91.

FROM THE HANDS OF A SLAVE

73 In 2008 Moody's rated greater Naples: See http://www.city-data.com/forum/business-finance-investing/314367-richest-cities-us-statistics.html.

74 Immokalee's per capita: Data from the U.S. Census Bureau: http://factfinder.census.gov/servlet/ACSSAFFFacts?_event=Search&geo_id=&_geoContext=&_street=&_county=immokalee&_cityTown=immokalee&_state=04000US12&_zip=&_lang=en&_sse=on&pctxt=fph&pgsl=010

75 Your chances of becoming a victim: From http://www.neighborhoodscout.com/fl/immokalee/crime/.

75 As a United States attorney: I interviewed Molloy on October 21, 2008.

75 From 2005 to 2007: Much of this information was obtained through court records related to *United States of America v. Cesar Navarrete, Geovanni Navarrete, Villhina Navarrete, Ismael Michael Navarrete, Antonia Zuniga Vargas*, United States District Court, Middle District of Florida, Fort Myers Division, Case no. 2:S07-cr-136-FtM-29DNF. On several occasions, I also interviewed members of the Coalition of Immokalee Workers who had knowledge of the case. U.S. Attorney Douglas Molloy and Collier County Sherriff Charlie Frost sat for lengthy interviews. Anyone writing about the conditions in Immokalee owes an enormous debt to Amy Bennett Williams for her ongoing coverage of a story that most Floridians did not know about. U.S. Attorney Molloy told me that slavery could not survive if the light of the media was shone upon it. No one has brightened that dark corner of our society more than Williams.

76 "The food was terrible": I interviewed Medel by telephone on November 15, 2010.

79 He allerted his colleague Charlie Frost: I interviewed Frost on October 21, 2008.

80 a total of about fifteen thousand: Due to the nature of the crime, human trafficking statistics are difficult to pin down. My source was the Polaris Project, http://nhtrc.polarisproject.org/images/nhtrcdocs/human-trafficking-statistics2.pdf. Data on murder rates came from the Federal Bureau of Investigation's Criminal Justice Information Services Division, http://www2.fbi.gov/ucr/cius2009/index.html.

82 Slavery and agriculture have had a close relationship in Florida: See "An Examination of the History and Evolution of Slavery in Florida's Fields," Florida Modern-Day Slavery Museum, http://www.ciw-online.org/freedom_march/MuseumBookletWeb.pdf.

83 tranquility was shattered: See John Bowe, *Nobodies: Modern American Slave Labor and the Dark Side of the New Global Economy* (New York: Random House, 2007), p. 44.

84 It required four more years: All information about the Flores case came from *United States of America v. Miguel A. Flores, Sebastian Gomez, Andres Ixcoy, and Nolasco Castaneda*, District Court of the United States, District of South Carolina, Charleston Division, Criminal case no. 2:96.806, October 10, 1996; also U.S. Department of Justice press release http://www.justice.gov/opa/pr/1997/November97/482cr.htm.html, United States Court of Appeals for the Fourth District, *United States of America v. Miguel A. Flores*, No. 98-4178, http://vlex.com/vid/us-v-flores-18328685; and Kevin Bales and Ron Soodalter, *The Slave Next Door: Human Trafficking and Slavery in America Today* (Berkeley: University of California Press, 2009), pp. 54–59.

85 Like Lucas Domingo: See Bales and Soodalter, pp. 49–50. Also see Department of Justice press release, http://www.justice.gov/opa/pr/1999/May/216cr.htm.

87 Laura Germino is a slender woman: I interviewed Germino numerous times between October 2008 and September 2010 in person and on the telephone. Our tour of Lake Placid took place on March 24, 2010.

89 Ariosto Roblero was a Guatemalan: For an excellent chronicling of the Ramos case see John Bowe, *Nobodies: Modern American Slave Labor and the Dark Side of the New Global Economy* (New York: Random House, 2007), pp. 3–77. See also *United States of America v. Ramiro Ramos, Juan Ramos, and Jose Ramos*, District Court of the United States, Southern District of Florida, case no. 01-14019-CR.

92 Jose Navarrete broke down: Amy Bennett Williams, "Five Plead Guilty in Immokalee Slavery Case," *Fort Myers News-Press*, September 3, 2008.

93 That day arrived: Amy Bennett Williams, "Immokalee Family Sentenced for Slavery," *Fort Myers News-Press*, December 20, 2008.

94 awarded a farmer: See Florida Fruit and Vegetable Association press release, http://www.ffva.com/iMISpublic/AM/Images/Layout_Assets/1508site_1508_20080325T143026/pdf%20library/distinguishedservice.pdf.

94 Viacava, the Navarretes' defense attorney: See Amy Bennett Williams, "Immokalee Family Sentenced for Slavery," *Fort Myers News-Press*, December 20, 2008.

95 According to testimony: For a reference to Orrin Hatch's efforts to remove "knowing or having reason to know," see written testimony of Lucas Benitez, Joint General Interest Hearing Regarding the Rights of Migrant Workers in the United States, Inter-American Commission on Human Rights Hearing. 122nd period of Sessions (March 3, 2005).

AN UNFAIR FIGHT

97 I met Geraldo Reyes: This meeting and interview occurred on October 20, 2008.

99 life expectancy of a migrant worker: Alberto Moreno, Migrant Health Fact Sheet, Oregon Department of Human Services (July 2010), http://www.oregon.gov/DHS/ph/omh/migrant/migranthealthfactsheet.pdf.

100 According to U.S. Labor Department figures: See "National Agricultural Workers Survey," U.S. Department of Labor Employment and Training Administration, http://www.doleta.gov/agworker/naws.cfm.

100 Reyes introduced me to a worker named Emilio Galindo: The interview with Galindo took place on March 24, 2010.

102 Leaning wearily against the railing: The events described in this paragraph and the next took place on October 20, 2008.

104 Pascuala Sanchez and her three children: Larry Hannan and Ryan Mills, "Grief Grips Immokalee," *Naples Daily News*, March 6, 2007; Katy Bishop and Ryan Mills, "$6M Settlement Reached in Deadly Immokalee Trailer Park Fire," *Naples Daily News*, September 12, 2007.

104 where one-quarter of the residences: Janine Zeitlin, "Not Giving Up on Fixing Up Immokalee Housing," *Naples Daily News*, October 22, 2006.

104 After touring Immokalee in 2008, Senator Bernie Sanders: See the transcript of the Hearing of the Committee on Health, Education, Labor and Pensions, United States Senate, One Hundred and Tenth Congress Second Session on Examining Abuses and Improving Working Conditions for Tomato Workers (April 15, 2008). http://frwebgate.access.gpo.gov/cgi-bin/getdoc.cgi?dbname=110_senate_hearing&docid=41-881.

105 attempt to shutter: Tracy X. Miguel, "Immokalee Migrant Workers Could Lose Their Homes Due to Collier Zoning Violations," *Naples Daily News*, January 18, 2009.

105 One of the reasons that rents: The explanation of why rents are so high in Immokalee and the description of the workers gathered at La Fiesta as being "the bottom of the bottom" of America's workforce came from an interview with Greg Asbed of the Coalition of Immokalee Workers, October 20, 2008.

107 One morning I encountered Lucas Benitez: The meeting and interview took place October 20, 2008. Benitez provided background on the coalition's early activities.

110 strikers began to weaken: Donald P. Baker, "Florida Farm Workers Fast for Better Wages," *Washington Post*, January 13, 1998.

111 Campaign for Fair Food: For details about the campaign, see the Coalition of Immokalee Workers' Web site, http://www.ciw-online.org.

111 grape boycotts mounted by Cesar Chavez: For an excellent account of Chavez's activities see Miriam Pawel, *The Union of Their Dreams: Power, Hope, and Struggle in Cesar Chavez's Farm Worker Movement* (New York: Bloomsbury Press, 2009).

112 "New Hedonism Generation": See http://www.ciw-online.org/tbnyoumatrix.html.

113 At a 2003 shareholders' meeting: See http://www.thefreelibrary.com/YUM+Brands+Shareholders+Demonstrate+Strong+Support+for+Proposal...-a0101935074.

114 Internet pseudonym surfxaholic36: Amy Bennett Williams, "Burger King Exec Uses Daughter's Online ID to Chide Immokalee Coalition," *Fort Myers News-Press*, April 28, 2008.

115 textbook-quality public relations flub: Elaine Walker, "Tomato Companies, Workers and Fast Food Firms Square Off," *Miami Herald*, November 20, 2007.

118 exiled Brazilian scholar: Paulo Freire, *Pedagogy of the Oppressed* (New York: Continuum, 2000).

120 $100,000 fine: See Steven Greenhouse, "Tomato Pickers' Wages Fight Faces Obstacles," the *New York Times*, December 24, 2007.

A PENNY PER POUND

121 Technically Brown has three jobs: I interviewed Brown on June 3, 2010.

122 I encountered a grower named Joe Procacci: I interviewed Procacci on March 2, 2005.

123 Agricultural Marketing Agreement Act of 1937: See http://www.ams.usda.gov/AMSv1.0/getfile?dDocName=STELPRDC5067868.

124 By taking on Procacci: For a company profile and history, see Doug Ohlemeier, "Procacci Bros. Marks 60 Years in Business," *The Packer*, January 9, 2001.

125 introduced legislation that would specifically exempt UglyRipes: See http://edocket.access.gpo.gov/2006/06-5833.htm.

126 Florida tomato growers have been falling further behind: John J. Van Sickle, "Spatial and Vertical Price Transmission in Fresh Produce Markets" (presented at the Agricultural Markets Workshop, April 21, 2006).

126 In the last three decades: I arrived at these figures by comparing the Consumer Price Index with the Producer Price Index for fresh-market, field-grown tomatoes as determined by the U.S. Department of Agriculture Economic Research Service. http://usda.mannlib.cornell.edu/MannUsda/viewDocumentInfo.do?documentID=1212.

127 a market that was awash: See Liam Pleven and Carolyn Cui, "Dying on the Vine: Tomato Prices—Tomatoes Go from Shortage to Glut in a Matter of Weeks," *Wall Street Journal*, June 17, 2010.

127 a massive salmonella outbreak: See Mickie Anderson, "UF Research Finds Salmonella Responds Differently to Varieties, Ripeness," *University of Florida News*, September 21, 2010.

128 Mexican imports accounted for about one-fifth: See "Vegetables and Melons: Tomatoes" briefing, U.S. Department of Agriculture, Economic Research Service (updated October 5, 2009).

128 Mexico agreed to a settlement: Eili Klein, "Rotten Tomatoes! The Mexican Growers Tomato Suspension Agreement and Its Effects on Mexico's Market Share: A Constant Market Shares Approach," Johns Hopkins School of International Studies, April 2004. http://www.princeton.edu/~eklein/pubs/MexicoMarketShareCMS_u2.pdf.

131 The skeptics' position was vindicated: See Elaine Walker, "Tomato Companies, Workers and Fast Food Firms Square Off," *Miami Herald*, November 20, 2007.

132 Brown was given an opportunity: For a transcript of the U.S. Senate committee hearing, see http://help.senate.gov/hearings/hearing/?id=0d03081e-0186-dc43-8fdf-0d68cfc2fc80.

135 Crist agreed to meet: Bill Maxwell, "Gov. Crist Backs Farmworkers," *St. Petersburg Times*, April 5, 2009.

137 Five months later: See Amy Bennett Williams, "Tomato Struggle Over after Immokalee Coalition Signs Historic Deal," *Fort Myers News-Press*, November 17, 2010.

MATTERS OF TASTE

140 I met John Warner Scott: I interviewed Scott in person on February 22, 2010. We also had several follow-up telephone conversations.

141 As for the reasonably fresh tomatoes: Thomas Whiteside, "Tomatoes," the *New Yorker*, January 24, 1977.

143 a tomato can lose its taste if exposed to cold temperatures: Trevor V. Suslow and Marita Cantwell, "Tomato Recommendations for Maintaining Postharvest Quality," University of California, Postharvest Technology Research and Information Center, February 10, 2009, http://postharvest.ucdavis.edu/Produce/ProduceFacts/Veg/tomato.shtml.

146 Harry Klee, a fellow University of Florida professor: I interviewed Klee in person on February 23, 2010. We also had several follow-up telephone conversations.

151 In consultation with Howard Moskowitz: For more about Moskowitz's work, see Malcolm Gladwell, "The Ketchup Conundrum," the *New Yorker*, September 6, 2004, http://www.gladwell.com/2004/2004_09_06_a_ketchup.html.

BUILDING A BETTER TOMATO

153 Tom Beddard had provided me: I interviewed Beddard on October 13, 2010.

157 Barbara Mainster cracked open a door: I interviewed Mainster on October 13, 2010.

159 In truth government grants: See Redlands Christian Migrant Association 2008–2009 Annual Report. http://www.rcma.org/annual%20report/RCMAannualrept2008-2009.pdf.

161 hired Steven Kirk to oversee reconstruction: I interviewed Kirk on October 14, 2010.

165 an umbrella organization called Rural Neighborhoods: See http://www.faqs.org/tax-exempt/FL/Rural-Neighborhoods-Incorporated.html#revenue_a.

TOMATOMAN

169 Tim Stark and I were in his pickup truck: I interviewed Stark on September 14 and 15, 2010.

170 in his 2008 memoir: Tim Stark, *Heirloom: Notes from an Accidental Tomato Farmer*, (New York: Broadway Books, 2008).

171 Stark told an NPR interviewer: See "Heirloom Tomato Farmer Finds Beauty in the Ugly, August 8, 2008," http://www.npr.org/templates/story/story.php?storyId=93356124.

178 Stark was still grumbling: Rob Patronite and Robin Raisfeld, *New York*, August 15, 2010, http://nymag.com/restaurants/features/67495/.

182 I was reminded of a passage: Michael Pollan, *The Omnivore's Dilemma: A Natural History of Four Meals* (New York: Penguin, 2006), pp. 249–50.

WILD THINGS

186 just outside Tembladera: For more information about the 2009 trip, see http://irbtomato.blogspot.com/.

AFTERWORD: NEW WORLD, OLD CHALLENGES

189 On a crisp, sunny spring morning: This encounter took place on April 14, 2011.

190 Fair Food Agreement: See http://www.ciw-online.org/FTGE_CIW_joint_release.html.

191 (since renamed Lipman): http://www.lipmanproduce.com/2011/09/six-l%E2%80%99s-packing-company-rebrands-changes-name-to-lipman

192 required to watch a video: see http://www.youtube.com/watch?v=UU4mPKxu_Uo&feature=player_embedded

194 My protest-marching skills: The demonstration took place on February 27, 2011.

196 A spokeswoman for the company told the *Boston Globe*: Stewart Bishop, "Stop & Shop protest march urges more pay for tomato farmers," *Boston Globe*, February 28, 2011.

196 In late 2011: The date was October 21, 2011.

197 Rev. Sarah Halverson, of Fairview Community Church: This interview was done over the telephone on October 26, 2011.

Bibliography

Allen, Arthur. *Ripe: The Search for the Perfect Tomato*. Berkeley, CA: Counterpoint, 2010.

Bales, Kevin, and Ron Soodalter. *The Slave Next Door: Human Trafficking and Slavery in America Today*. Berkeley: University of California Press, 2009.

Barndt, Deborah. *Tangled Routes: Women, Work, and Globalization on the Tomato Trail*. 2nd ed. Lanham, MD: Rowman and Littlefield, 2008.

Bowe, John. *Nobodies: Modern American Slave Labor and the Dark Side of the New Global Economy*. New York: Random House, 2007.

Conover, Ted. *Coyotes: A Journey through the Secret World of America's Illegal Aliens*. New York: Vintage, 1987.

Fromartz, Samuel. *Organic, Inc.: Natural Foods and How They Grew*. Orlando: Harcourt, 2006.

Goldman, Amy. *The Heirloom Tomato from Garden to Table: Recipes, Portraits, and History of the World's Most Beautiful Fruit*. New York: Bloomsbury, 2008.

Kimbrell, Andrew, ed. *The Fatal Harvest Reader: The Tragedy of Industrial Agriculture*. Washington, DC: Island Press, 2002.

Livingston, Alexander W. *Livingston and the Tomato*. Foreword and appendix by Andrew F. Smith. Columbus: Ohio State University Press, 1998.

Martineau, Belinda. *First Fruit: The Creation of the Flavr Savr Tomato and the Birth of Genetically Engineered Food*. New York: McGraw-Hill, 2001.

Martínez, Rubén. *Crossing Over: A Mexican Family on the Migrant Trail*. New York: Metropolitan Books, 2001.

McPhee, John. *Oranges*. New York: Farrar, Straus & Giroux, 1967.

Olson, Stephen M., and Bielinski Santos, eds. *Vegetable Production Handbook for Florida 2010–2011*. Gainesville: University of Florida, 2010.

Parsons, Russ. *How to Pick a Peach: The Search for Flavor from Farm to Table*. Boston: Houghton Mifflin, 2007.

Pawel, Miriam. *The Union of Their Dreams: Power, Hope, and Struggle in Cesar Chavez's Farm Worker Movement*. New York: Bloomsbury Press, 2009.

Pawlick, Thomas F. *The End of Food: How the Food Industry Is Destroying Our Food Supply—and What You Can Do about It*. Fort Lee, NJ: Barricade Books, 2006.

Pollan, Michael. *The Omnivore's Dilemma: A Natural History of Four Meals.* New York: Penguin, 2006.

Rothenberg, Daniel. *With These Hands: The Hidden World of Migrant Farmworkers Today*. Berkeley: University of California Press, 2000.

Schlosser, Eric. *Fast Food Nation: The Dark Side of the All-American Meal.* Boston: Houghton Mifflin, 2001.

Schlosser, Eric. *Reefer Madness: Sex, Drugs, and Cheap Labor in the American Black Market.* Boston: Houghton Mifflin, 2003.

Smith, Andrew F. *The Tomato in America: Early History, Culture, and Cookery*. Columbia: University of South Carolina Press, 1994.

Stark, Tim. *Heirloom: Notes from an Accidental Tomato Farmer*. New York: Broadway Books, 2008.

Stewart, Keith. *It's a Long Road to Tomato: Tales of an Organic Farmer Who Quit the Big City for the (Not So) Simple Life*. New York: Marlowe, 2006.

Thissen, Carlene A. *Immokalee's Fields of Hope*. New York: iUniverse, 2004.

Watt, Bernice K., and Annabel L. Merrill. *Composition of Foods: Raw, Processed, Prepared*, Agricultural Handbook No. 8. Washington, DC: United States Department of Agriculture, 1964.

Wilkinson, Alec. *Big Sugar: Seasons in the Cane Fields of Florida*. New York: Knopf, 1989.

Index

C

D

G

I

J

K

L

M

N

O

P

R

T

Y

Z

Photo by Trent Campbell

James Beard Award–winning journalist Barry Estabrook was a contributing editor at *Gourmet* magazine for eight years, writing investigative articles about where food comes from. He was the founding editor of *Eating Well* magazine and has written for the *New York Times Magazine*, *Reader's Digest*, *Men's Health*, *Audubon*, and the *Washington Post*, and contributes regularly to *The Atlantic*'s Web site. His work has been anthologized in the Best American Food Writing series, and he has been interviewed on numerous television and radio shows. He lives and grows tomatoes in his garden in Vermont. Visit him at www.politicsoftheplate.com.